Getting Started with Netduino

Chris Walker

O'REILLY®

Beijing · Cambridge · Farnham · Köln · Sebastopol · Tokyo

Getting Started with Netduino
by Chris Walker

Published by O'Reilly Media, Inc., 1005 Gravenstein Highway North, Sebastopol, CA 95472.

O'Reilly books may be purchased for educational, business, or sales promotional use. Online editions are also available for most titles (*http://my.safaribooksonline.com*). For more information, contact our corporate/institutional sales department: (800) 998-9938 or *corporate@oreilly.com*.

Editor: Brian Jepson
Production Editor: Kristen Borg
Proofreader: O'Reilly Production Services
Cover Designer: Karen Montgomery
Interior Designer: Ron Bilodeau and Edie Freedman
Illustrator: Marc de Vinck

February 2012: First Edition.

Revision History for the First Edition:
February 9, 2012   First release
See *http://oreilly.com/catalog/errata.csp?isbn=9781449302450* for release details.

ISBN: 978-1-449-30245-0
[LSI]
1328800561

This book is dedicated to the Netduino Community:

Tens of thousands of tinkerers have picked up Netduino and started tinkering.

So many have shared their experiences, projects, and friendship.

Thank you for making our journey so fun and inspiring.

Contents

Preface

Computers surround us. I'm not speaking of laptops or tablets or cell phones. Billions of remote controls, thermostats, sensors, and gadgets of all sorts have little computers inside them. And while millions of software engineers develop applications for phones, computers, and the Web—the programming languages and skills that apply there are quite a bit different than those needed to develop code for tiny embedded microcontrollers.

Or at least, they were.

In 2004, Microsoft introduced the SPOT Smart Watch. It ran a tiny version of their desktop .NET programming runtime and enabled application developers to write software for its tiny microcontroller using the C# programming language they already knew. Almost a decade later, this runtime is now in its fourth major version, is running on millions of devices around the world, and has grown to enable tinkerers to use traditional software development skills to build their own electronics projects with Netduino.

Like me, you may be a tinkerer. You may like building things or tearing things apart to understand how they work. You may want to build your own web-based coffee machine, Morse code generator, or electronically enhanced Halloween costume.

Or you may be an educator or student who wants to learn how electronics work. Netduino and the .NET Micro Framework enable you do this without drowning in a sea of datasheets, and without needing to understand the intricacies of microcontroller registers at the same time.

Because Netduino is open source, all design files and source code are included. If you desire to become an expert or just need a reference to understand how things work behind the scenes, that is all provided at no charge. Netduino gives you freedom to build, to remix, and to have fun.

And for many, Netduino is about fun: to learn how electronics work, to build cool projects, and to play. For others, Netduino is a serious tool used to develop viable products.

Whether you are interested in Netduino for fun or for profit, there is an online community of tens of thousands of fellow makers at *http://forums .netduino.com*. Come join us in learning, in building, and in sharing our electronics achievements. I look forward to meeting you there.

What You Need to Know

This book is written with the goal of giving non-programmers enough training to successfully build the samples in this book, while providing software engineers the ability to delve into electronics using their sophisticated programming skills.

If you have written scripts for a web page, you have the skills necessary to tackle this book. If you have used a word processor before, you should be able to follow along with the samples. And if you write desktop or web applications in C# for a living, you should enjoy this journey.

How to Use This Book

This book is intended to be read from beginning to end. I start by introducing Netduino and walking you through installation of the free Windows-based development tools. I then introduce electronic components like buttons, LEDs, and speakers—and show you how to use them. Finally, I show you how to connect an Internet-enabled Netduino to the Web.

These examples are all building blocks, giving you tools in your electronics tool chest. And while it is a lot of fun to write code that changes the color of LEDs or plays music on a tiny speaker, the building blocks are ultimately for use in your own projects.

As you build your own projects or reproduce the projects of fellow Netduino community members, you'll inevitably need electronics parts. Stores like RadioShack and Micro Center can provide you with drawers full of components you can use to build projects, and online stores like MakerShed (*http://www.makershed.com*) provide both parts and full kits (with instructions) that can guide you in your tinkering ways.

Enjoy the journey. It's a lot of fun.

Conventions Used in This Book

The following typographical conventions are used in this book:

Italic
> Indicates new terms, URLs, email addresses, filenames, and file extensions.

`Constant width`
> Used for program listings, as well as within paragraphs to refer to program elements such as variable or function names, databases, data types, environment variables, statements, and keywords.

Constant width bold
> Shows commands or other text that should be typed literally by the user.

Constant width italic
> Shows text that should be replaced with user-supplied values or by values determined by context.

 TIP: This icon signifies a tip, suggestion, or general note.

Using Code Examples

This book is here to help you get your job done. In general, you may use the code in this book in your programs and documentation. You do not need to contact us for permission unless you're reproducing a significant portion of the code. For example, writing a program that uses several chunks of code from this book does not require permission. Selling or distributing a CD-ROM of examples from O'Reilly books does require permission. Answering a question by citing this book and quoting example code does not require permission. Incorporating a significant amount of example code from this book into your product's documentation does require permission.

We appreciate, but do not require, attribution. An attribution usually includes the title, author, publisher, and ISBN. For example: "*Getting Started with Netduino* by Chris Walker (O'Reilly). Copyright 2012 Secret Labs LLC, 978-1-4493-0245-0."

If you feel your use of code examples falls outside fair use or the permission given above, feel free to contact us at *permissions@oreilly.com*.

Safari® Books Online

 Safari Books Online is an on-demand digital library that lets you easily search over 7,500 technology and creative reference books and videos to find the answers you need quickly.

With a subscription, you can read any page and watch any video from our library online. Read books on your cell phone and mobile devices. Access new titles before they are available for print, and get exclusive access to manuscripts in development and post feedback for the authors. Copy and paste code samples, organize your favorites, download chapters, bookmark key sections, create notes, print out pages, and benefit from tons of other time-saving features.

O'Reilly Media has uploaded this book to the Safari Books Online service. To have full digital access to this book and others on similar topics from O'Reilly and other publishers, sign up for free at *http://my.safaribooksonline.com*.

How to Contact Us

Please address comments and questions concerning this book to the publisher:

O'Reilly Media, Inc.
1005 Gravenstein Highway North
Sebastopol, CA 95472
800-998-9938 (in the United States or Canada)
707-829-0515 (international or local)
707-829-0104 (fax)

We have a web page for this book, where we list errata, examples, and any additional information. You can access this page at:

http://shop.oreilly.com/product/0636920018032.do

To comment or ask technical questions about this book, send email to:

bookquestions@oreilly.com

For more information about our books, courses, conferences, and news, see our website at *http://www.oreilly.com*.

Find us on Facebook: *http://facebook.com/oreilly*

Follow us on Twitter: *http://twitter.com/oreillymedia*

Watch us on YouTube: *http://www.youtube.com/oreillymedia*

This Book Was Made Possible By

In the 18 months since the launch of Netduino, so much has happened...and the fun has only started.

Like Netduino itself, this book couldn't have happened without the help of so many others. Here are just a few of the people that made this all possible.

Thank you to:

You—for purchasing this book and joining the growing worldwide community of Netduino tinkerers. You are why Netduino exists.

Brian Jepson, Limor Fried, and Phillip Torrone—who taught me the importance of open source hardware and have been so supportive of our efforts.

Clint Rutkas and Microsoft's Channel 9 Team—who have celebrated many of the community's awesome projects, and who make hacking electronics fun!

Marc de Vinck and the MakerShed team—who were the first to take a chance on Netduino and bring it to a larger audience online.

Stefan Thoolen—for spending dozens and dozens of hours reviewing the book, making recommendations for sample code, creating the netmftoolbox, and more. Also, for being the best moderator ever. You are a good friend.

The Arduino team, especially Massimo Banzi and Tom Igoe—for kickstarting the open hardware movement, for continuing to introduce so many people to electronics, and for warmly inviting me into their world.

Brian Jepson and Marc de Vinck—thank you a second time—you have been so patient during the making of this book, and then you edited and illustrated it so nicely. You're the best.

Colin Miller, Lorenzo Tessiore, and Zach Libby from Microsoft—thank you for giving life to .NET Micro Framework, for open sourcing it, and for continuing to develop the core platform all these years.

David Stetz, Stanislav Simicek, fellow .NET Micro Framework core tech team members, and others who have contributed to the codebase—your contributions are the lifeblood of .NET Micro Framework.

Miguel de Icaza and the Mono team—for making it possible for Mac and Linux users to write code for Netduino too, and for your open source passion.

Cuno Pfister—your book *Getting Started with the Internet of Things* was the catalyst for this book, and all your contributions to the community are greatly appreciated.

Secret Labs staff—for demanding perfection and for being so sneaky and quiet while I finished this book.

The entire O'Reilly team—for believing in Netduino and for making this book possible.

1/Introducing Netduino

Netduino is an electronics platform. Using Netduino, hobbyists and pro-grammers can create electronics projects (and electronics-based art projects) with ease. Several Netduino boards are available, which I explore in detail later in this chapter.

Netduino apps use the *.NET Micro Framework*. This programming frame-work from Microsoft is easy to get started with; for beginners, writing .NET code is as easy as creating simple JavaScript animations for a web page. And for the millions of programmers who already write .NET code, the .NET Micro Framework provides an incredibly powerful set of features (such as events, threading, and line-by-line debugging).

Traditional microcontroller code consists of a fixed loop where code checks conditions and performs actions based on those conditions. In contrast, the .NET Micro Framework empowers you to break your app's tasks into simpler routines. You can execute those routines after a certain amount of time or in response to hardware actions. You can even multitask! These fea-tures let you build very sophisticated apps using easy-to-understand code.

Microsoft provides free software tools to create Netduino apps. All you need to get started is a Netduino, a computer, and your imagination. As you pro-gress through the projects in this book, I'll introduce expansion shields and electrical components that you can add to your Netduino to create larger projects.

 TIP: If you use a Mac or a Linux computer, the Mono project provides an alternative set of .NET pro-gramming tools. While the examples in this book use Microsoft's Visual Studio Express tools, Ap-pendix B explains how you can program a Netduino using Mono on a Mac or Linux computer.

Finally, the Netduino community is made up of tens of thousands of tinkers, many of whom share their projects and help each other out on the Netduino community forums (*http://forums.netduino.com*).

As you get started with Netduino, I invite you to join this growing community.

 TIP: Netduino is also open source. This means that the source code for the Netduino firmware is provided (Apache 2.0 and BSD open source licenses). The hardware design files—including schematics and engineering layout files—are also provided (Creative Commons-Attribution license). Many engineers will prototype new electronics products using Netduino—and then create custom hardware designs based on these open source files. And many software programmers will add to the source code to create new features for Netduino—and then share those enhancements with friends.

Meet the Netduino Family

The Netduino family consists of three electronics boards: Netduino, Netduino Plus, and Netduino Mini.

Netduino (shown in Figure 1-1) is the entry-level board in the Netduino family.

The Netduino board is made up of several electronic components and connectors, most notably:

Atmel ARM microcontroller
> This is the main processor, and it contains the code storage space and RAM used by your Netduino app. The microcontroller's pins are wired to the blue pin headers, enabling your Netduino app to connect to external components and expansion shields.

6 Analog Input headers
> - You can plug sensors into these headers (light, temperature, motion, pressure, etc.).
> - Analog Input headers can also operate in digital I/O mode (explained next).

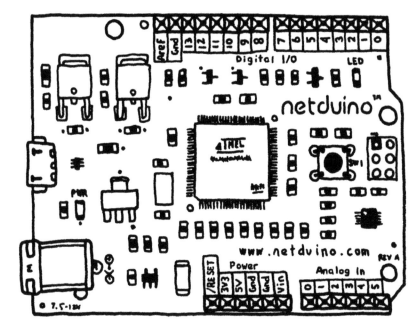

Figure 1-1. *Netduino*

14 Digital Input/Output headers

- You can plug on/off inputs into these headers (switches, pushbuttons, on/off sensors).
- You can plug on/off outputs into these headers (LEDs, relay switches, etc.).
- Several Digital I/O headers can communicate using standard communication protocols: I2C, SPI, UART (serial).
- Some Digital I/O headers can pulse electricity to change the speed of motors, control the intensity of LEDs, and more.

Pushbutton

- By default, the pushbutton resets the Netduino and restarts your Netduino app. This can be useful to restart the sequence of actions taken by your app.
- Alternatively, the pushbutton can be used as an input: when pressed, your app can then take various actions.
- Finally, holding down the pushbutton while powering up your Netduino will temporarily put the microcontroller into programming mode (for firmware updates).

Power and user LEDs

- The white power LED is illuminated while the Netduino is powered.
- The blue user LED turns on briefly when the Netduino is first powered and then shuts off to let you know that the board has booted. You can turn this LED on and off in your Netduino app.

Power barrel jack

- The Netduino may be powered by an AC-to-DC power adapter with a standard 5.5mm (outer)/2.1mm (inner) plug. Allowable voltages are from 7.5V to 12V, and the plug polarity must be center pole positive.
- To help protect you in case you accidentally connect a "center negative"adapter instead of a "center positive" adapter, the board has an integrated reverse voltage protection fuse. This fuse will auto-reset once it cools down.

Power regulation circuitry

Onboard power regulators convert incoming higher voltage into the 3.3V needed by the microcontroller. They also provide power to the 5V and 3.3V pin headers, for use by external components and expansion shields.

MicroUSB port

- The MicroUSB port connects the Netduino to your computer's USB port.
- You can also power the Netduino from your computer (or USB power supply) over the MicroUSB port.
- By default, the MicroUSB connection is used to deploy apps to your Netduino and to interactively debug those apps.
- Alternatively, the MicroUSB connection can be used to turn your Netduino into a USB device (keyboard, mouse, bi-directional communication device, etc.).

Erase pad

- The erase pad is a small gold square located directly underneath digital pin 0.
- By connecting a wire between the 3.3V header and the erase pad, you will erase your Netduino completely. Netduino is designed to be hacker-friendly and the erase pad lets you start over from scratch or repurpose your Netduino as an ARM microcontroller development board.

- By erasing your Netduino, you can install alternative operating systems, write native C++ code on your Netduino, flash the Netduino firmware from scratch, and so forth.

Netduino Plus (see Figure 1-2) is an enhanced version of the Netduino, adding storage and networking features.

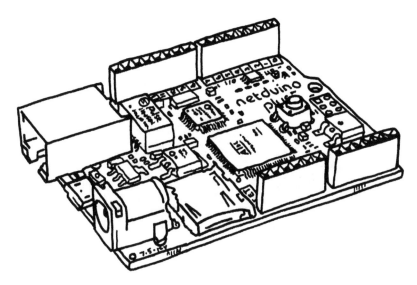

Figure 1-2. *Netduino Plus*

The Netduino Plus board adds the following components:

Ethernet jack

- The Ethernet jack allows your Netduino Plus to connect to the Internet through a network router.
- After connecting your Netduino Plus to the Internet, you can post sensor data to Twitter, interact with a mobile phone app, communicate with far-away Netduinos, etc.

MicroSD slot

- The MicroSD slot lets you add persistent storage to your Netduino app. You can log environmental data, store web pages (to serve to the Internet), and much more.
- Advanced users can store compiled code on MicroSD cards, for execution by the Netduino Plus or a Netduino with an add-on card reader.

Netduino Mini (Figure 1-3) is basically a tiny, scaled down, breadboard-friendly version of the Netduino:

- Netduino Mini is 0.72 square inches (4.65 cm^2) small.
- Netduino Mini has four analog pins instead of six, and has two fewer digital pins than Netduino.
- Netduino Mini is designed for industrial temperature ranges (-40 to 185F, -40 to 85C).
- Netduino Mini is programmed via a serial cable (RS232 or TTL) instead of via USB.

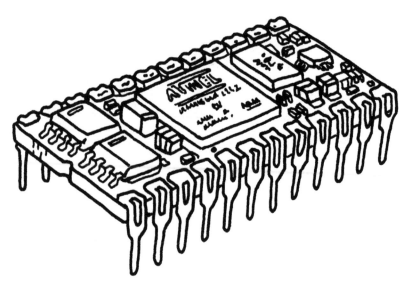

Figure 1-3. *Netduino Mini*

What You Need

To build the projects in the following chapters, you'll need either a Netduino or Netduino Plus. If you have a Netduino Mini, most of these samples will work—but you'll need to adjust the analog and digital pin numbers in your code to reflect the Netduino Mini's form factor.

NOTE: Please note that Chapter 8 explores Internet-connected apps using Netduino Plus (which has a built-in Ethernet jack). It is also possible to add components to a "regular" Netduino or a Netduino Mini and connect to the Internet; refer to the online community for guidance on how to do that.

You'll learn a few programming fundamentals as you progress through the chapters. Experienced programmers can skip those paragraphs if they so desire. My goal is to ensure that all readers gain the knowledge to successfully build their own electronics projects.

2/Setting up the Free Tools

Netduino apps are built using Microsoft's C# programming language. Straightforward yet powerful, C# allows us to focus on the "what" and "when" of programming, while letting the Netduino take care of all the low-level details.

Microsoft creates a world-class code editor (with a rich development environment) that you can use to write and debug Netduino apps. This editor is named Visual Studio, and there are both free and paid versions available. You'll use the free version (Visual Studio Express) for the examples in this book; for programmers who already own a high-end version of Visual Studio 2010, you can safely skip to "Step 2: Install the .NET Micro Framework SDK" on page 12.

Apps written using C# execute within the *Common Language Runtime* (CLR). This Runtime provides the rich feature set used by C# apps. It works with other programming languages as well—but for the purposes of this book, I focus on C#.

Netduino uses a custom runtime designed especially for microcontrollers: the .NET Micro Framework. This scaled-down runtime needs only a small amount of storage space and working memory, unlike the higher-end editions that run on phones and computers. It also includes a number of features specific to development boards like Netduino, such as the ability to input and output electronic signals and values.

To create Netduino apps, you need to install three software packages on your computer:

1. The Visual Studio development environment.
2. The .NET Micro Framework software development kit (SDK).
3. The Netduino software development kit (SDK).

TIP: At the time of writing: Visual Studio 2010 was the latest edition of the development environment; .NET Micro Framework v4.1 was the latest edition of the runtime; Netduino v4.1 SDK was the latest version of the Netduino SDK. Newer versions may be available when you read this, so be sure to grab the latest and greatest from the Netduino downloads page at *http://www.netduino.com/downloads/*.

Step 1: Install Visual Studio Express

Before you install any of the .NET Micro Framework or Netduino tools, install Visual Studio. At the time of writing, the download URL for the free version of Visual Studio is *http://www.microsoft.com/express/downloads/*.

On the download page, select the "Visual C# 2010 Express" option. You may also need to select your language. Download and run the web installer, then go through the following steps:

1. Click Next to go past the welcome screen, then review the license terms on the next screen.

TIP: While Visual Studio is the name of the development environment, Microsoft often calls the C#-specific version "Visual C# Express." You can consider both names to mean the same thing for our purposes.

2. After you accept the license terms, press Next.
3. At this step, you may optionally install additional software packages (Figure 2-1). None of these products are required to write apps for your Netduino.

   After selecting any optional products that you would like to install, press Next.
4. If desired, you may change the installation location of Visual Studio Express, as shown in Figure 2-2. If you don't have a good reason to change the default location, leave it as-is.

Figure 2-1. *Optional installation packages*

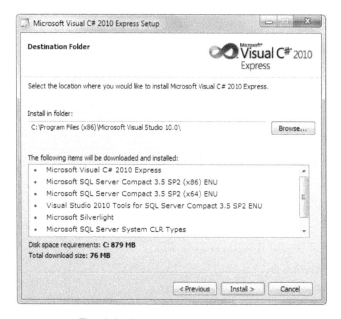

Figure 2-2. *The default installation location for Visual Studio*

Press Install to start the installation process. This will download all required files and install them on your computer. The process will take somewhere between 15 minutes and an hour, depending on the speed of your computer, the speed of your Internet connection, and how many optional components you selected.

Step 2: Install the .NET Micro Framework SDK

Now that we have installed the Visual Studio code editor, you need to install the .NET Micro Framework tools. These will allow you to write code for the .NET Micro Framework runtime and also install diagnostics tools, which we'll explore later.

Download the .NET Micro Framework SDK from the Netduino downloads page: *http://www.netduino.com/downloads/*. Once downloaded, double-click on the .NET Micro Framework SDK installer to start the installation process:

1. Once the installer starts up, click Next. After you accept the license terms, click Next.

2. The next step asks you to choose a setup type (see Figure 2-3). Select "Typical" and click Next.

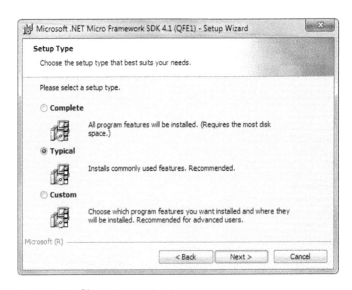

Figure 2-3. *Choosing a setup type*

3. The installer will ask you to confirm that you want to proceed. Click Install.

Step 3: Install the Netduino SDK

Now that you have the Visual Studio code editor and .NET Micro Framework software development tools installed, you must install the Netduino SDK. The Netduino SDK includes project templates that make it easy to get started with Netduino, USB drivers for Netduino, and other Netduino-specific tools.

Download the Netduino SDK from the Netduino downloads page: *http://www.netduino.com/downloads/*.

You may need to select between 32-bit and 64-bit options. Pick the one that matches your edition of Windows. If you don't know which to pick, click on the Start Button to bring up your Windows Start Menu. Then right-click on Computer and select the Properties menu option. A window will pop up with information about your computer, including the System Type (32-bit or 64-bit).

Once downloaded, double-click on the Netduino SDK installer to start the installation process:

1. After the installer starts up, click Next. After you accept the license terms, click Next.
2. If desired, you may change the installation location of Netduino SDK. If you don't have a good reason to change the default location, leave it as-is and click Next.
3. Click Install to start the installation process.
4. If Windows prompts you for permission to install the Netduino USB drivers, give it permission to do so.

Conclusion

You have finished installing the necessary tools to develop Netduino apps. Now it is time to write some code and make things with your Netduino.

3/First Projects

Microcontroller boards like Netduino live in a realm that bridges software and electronic hardware. The Netduino itself is the bridge, and your Netduino app determines how software and hardware talk to each other. This means that one of the fundamental activities you'll engage in is *signaling*: your app must send signals over digital outputs and receive signals on digital inputs.

These digital signals are represented in your app's code as binary 1s and 0s. Usually, 1s are represented by a higher voltage across a wire (on Netduino, that's 3.3 or 5 volts) and 0s are represented by a low voltage across a wire (on Netduino, 0 volts).

Using these 1s and 0s, you can send high and low voltages to electronic components like LEDs and relays (turning them on and off). You can also receive high and low voltages from components such as pushbuttons: when your app receives a 1, it means the button is pushed.

Both Netduino and Netduino Plus have an onboard LED that can be controlled from code. As a first project, you'll learn how to send digital 1s and 0s by blinking this LED. Later in this chapter, you'll learn how to read the state of Netduino's onboard pushbutton.

Start Visual Studio

Start Visual C# Express 2010 (if you use the full version of Visual Studio 2010, start Visual Studio 2010 instead). The installer (see "Step 1: Install Visual Studio Express" on page 10) created a folder and shortcut for this program in your Start menu under All Programs (Windows Vista or Windows 7) or Programs (Windows XP).

The Visual Studio programming environment will start. Now you're ready to create your first project:

1. Click on the New Project link. If no link is visible, go to the File menu and select New Project.

2. The New Project window should now pop up. Visual Studio displays a set of installed templates. We want to pick Visual C#→Micro Framework from the list on the left. Then pick Netduino Application from the list on the right. Give your project a name such as Blinky (as shown in Figure 3-1), and click OK.

 NOTE: If you haven't unpacked your Netduino, do so now. Attach its sticky feet. Grab a Micro USB cable and plug the Netduino into your computer. If you did not get a Micro USB cable with your Netduino, you may be able to borrow one from a cell phone or another device.

Figure 3-1. *Create a new project*

Blinking the Onboard LED

The first thing that microcontroller programmers often do is blink an LED on their new electronics board. This verifies that the board is booting up properly and that they can successfully create and run a simple app on the board.

While this can often take hours or days in the traditional microcontroller world, you'll do it in a few minutes. Go ahead and write this Blinky app now:

1. Now that Visual Studio is open, look for Solution Explorer on the right side of the screen. Solution Explorer shows the source and data files, which make up your Netduino project. Of particular note, the file *Program.cs* holds the startup code for your project. You're going to open it and write about a half dozen lines of code. Double-click on *Program.cs* now (or right-click on its name and select Open).

2. In the main section of the Visual Studio editor, you are now editing *Program.cs*. Click on the line underneath the text `// write your code here`. This is where you'll write your code.

3. Now, type the following:

```
OutputPort led = new OutputPort(Pins.ONBOARD_LED, false);
```

   This first line of code creates an `OutputPort`. An `OutputPort` is a software object that lets you control the voltage level of a *pin* on the Netduino. The first parameter tells the Netduino which pin of the microcontroller you want to control, and the second parameter tells the Netduino which state to put it in. `Pins.ONBOARD_LED` is shorthand that specifies the Netduino's built-in blue LED. The second parameter (false) puts the LED in an initial state of OFF (false).

 NOTE: For digital inputs and outputs, higher voltage (1) is represented as **true** and lower voltage (0) is represented as **false**.

---

4. Now, you're going to blink the LED on and off repeatedly. A straightforward way to create an action that repeats forever is to put it inside a loop that never ends. Add the following code to your project:

```
while (true)
{
}
```

   The keyword *while* tells the microcontroller to do something in a loop while a certain condition is met. This condition is placed in parentheses. In our case, we use a condition of **true**. Since conditions are met when they are true, putting **true** here means that the loop will repeat forever.

5. Next, create the blinking LED code. Between the two sets of curly braces, insert the following four lines of code:

```
led.Write(true); // turn on the LED
Thread.Sleep(250); // sleep for 250ms
led.Write(false); // turn off the LED
Thread.Sleep(250); // sleep for 250ms
```

Your final program's `Main()` method should now look like the following listing. Figure 3-2 shows Visual Studio Express with the Solution Explorer to the right and the complete *Program.cs* open in the editor:

```csharp
public static void Main()
{
    // write your code here
    OutputPort led = new OutputPort(Pins.ONBOARD_LED, false);

    while (true)
    {
        led.Write(true); // turn on the LED
        Thread.Sleep(250); // sleep for 250ms
        led.Write(false); // turn off the LED
        Thread.Sleep(250); // sleep for 250ms
    }
}
```

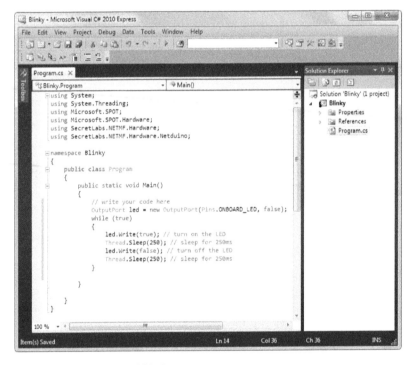

Figure 3-2. *The finished Blinky program*

Running the Blinky App

Next, you'll deploy the Netduino app to the Netduino and watch it run.

By default, Visual Studio runs projects in an *emulator*. This allows software developers to create and test programming logic for a new hardware product before the actual hardware is built. You won't use the emulator since you have a real Netduino, so you must let Visual Studio know that you have physical hardware it should use instead.

Click on the Project menu and select your project's properties (generally, the last item in the Projects menu, such as Blinky Properties). Next, do the following:

1. When the project properties appear, click on the .NET Micro Framework category on the left side.

2. Now you're ready to change the deployment target from the Emulator to the Netduino. Change the Transport from Emulator to USB and then make sure that the Device selection box shows your Netduino, as shown in Figure 3-3. If it doesn't, unplug and reattach your Netduino.

NOTE: If you're using a Netduino Plus, the target name will be different. So if you need to switch between deploying to a Netduino and Netduino Plus, you'll need to return to the project properties and change the Device setting to the device you want to deploy to.

Now, you're ready to run the project. When you run the project, your code is deployed to the Netduino and then automatically started. To run your project, click the Start Debugging button in the toolbar at the top of the screen. It looks like the Play button on a music player. You could also press its keyboard shortcut, F5.

NOTE: You'll just watch the program run for now, but when you start building sophisticated Netduino Apps, you may want to pause or step through your code line-by-line, peek at the values of data in the Netduino's memory, and so forth. If you want to go through the code line-by-line, one way to do that is to start the program with Step Into (F11) instead of Start Debugging.

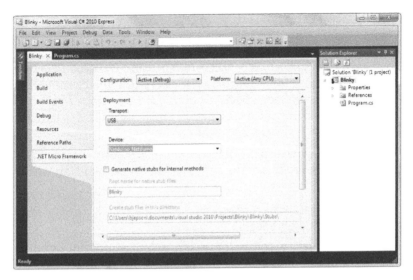

Figure 3-3. *Setting the project's .NET Micro Framework properties*

Visual Studio will now deploy your first Netduino app to the Netduino hardware. In a few seconds, you'll see the blue LED (Figure 3-4) blinking on and off every half second.

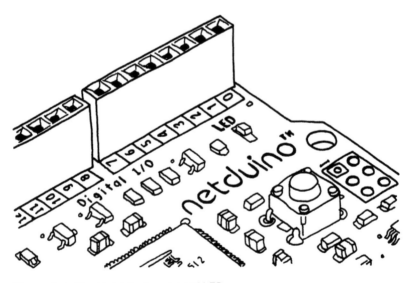

Figure 3-4. *The Netduino's onboard LED*

NOTE: When you started your Netduino app, it was written into the Netduino's microcontroller chip's flash memory, so all you have to do to run the program again is plug it in via a MicroUSB cable or with a power adapter (using the power barrel jack). You can write over your Netduino App. Visual Studio will automatically stop and erase your current Netduino App whenever deploying a new one.

Now that you've successfully written a simple Netduino app and controlled an LED by sending it digital output of high (true) and low (false) voltages, you're ready to learn how to read digital input signals.

Pushing the Onboard Button

Both Netduino and Netduino Plus have an on onboard pushbutton labeled SW1. You can read the digital value of this pushbutton—true or false—to determine whether it is currently being pushed. Then you can take actions based on that pushbutton's current state, such as turning an LED on and off or instructing a Netduino-powered robot to start moving its motors.

NOTE: By default, this pushbutton will reboot your Netduino and restart your current Netduino app. But because of a special configuration in the Netduino's circuitry, it can alternatively be used as a digital input.

In Visual Studio, create a new project by selecting New Project in the File menu. Select Netduino Application as the template as before, and give your project a name such as "Pushbutton1." Finally, double-click on *Program.cs* in the Solution Explorer to begin editing it.

In the main section of the Visual Studio editor, you are again editing *Program.cs*. Click on the line underneath the text // write your code here.

Now, type the following:

```
OutputPort led = new OutputPort(Pins.ONBOARD_LED, false);❶
InputPort button =
  new InputPort(Pins.ONBOARD_SW1, false, Port.ResistorMode.Disabled);❷
bool buttonState = false;❸
```

❶ As with the previous example, the first line of code creates an `Output Port` so that you can turn the onboard LED on and off.

❷ The second line of code creates an *InputPort*. An InputPort lets you read the voltage level of the pins on the Netduino (or in this case, the voltage coming from the pushbutton). `Pins.ONBOARD_SW1` is shorthand that tells the Netduino which pin of the microcontroller to use for input. The second value, false, tells the runtime that you don't need glitch filtering (if you enable it, Netduino will take multiple readings during a button press to make sure the reading is correct). The final value, `Port.ResistorMode .Disabled` indicates that Netduino won't use a built-in resistor to affect incoming signals on the microcontroller's digital pin.

 NOTE: I explore the resistor mode in greater detail later in "Pushing the MakerShield's Button" on page 34.

---

❸ The third line of code creates a variable named `buttonState`. A variable is a way to store and manipulate data in the memory of the Netduino. In this case, the app stores whether or not the Netduino's pushbutton is currently being pushed in the `buttonState` variable. The word *bool*, which precedes the variable, indicates that `buttonState` will store a Boolean value. Boolean values are values that are true or false, perfect for this application. Finally, `= false` sets the state of the value to false, by default. It's generally a good idea to set variables to a default value so that you make your intentions clear to anyone else (including you!) who reads the source code in the future.

Now you're going to read the state of the pushbutton and turn the LED on and off as the pushbutton is pushed and released. First, create an infinite loop as before. Add the following code to your project:

```
while (true)
{
}
```

Then, read the current state of the pushbutton and write out that state to the LED. Between the two sets of curly braces, insert the following two lines of code:

```
buttonState = button.Read();
led.Write(buttonState);
```

Your final program's **Main()** method should look like this:

```
public static void Main()
{
    // write your code here
    OutputPort led = new OutputPort(Pins.ONBOARD_LED, false);
    InputPort button =
        new InputPort(Pins.ONBOARD_SW1, false, Port.ResistorMode.Disabled);

    bool buttonState = false;

    while (true)
    {
        buttonState = button.Read();
        led.Write(buttonState);
    }
}
```

Now, you're almost ready to run the app, press the pushbutton, and use it to control the state of the LED. But first, you need to make sure you're deploying the project to the Netduino instead of to the built-in emulator. Click on the Project menu and select your project's properties. Then click on the .NET Micro Framework category on the left side. Change the Transport from Emulator to USB, and then make sure that the Device selection box shows your Netduino.

Now run your project. Press the Start Debugging button in the toolbar at the top of the screen or press F5.

After a few seconds, your Netduino app will be running. Once the blue LED turns off, you know that your board has booted. Press the pushbutton, and the blue LED will turn on. Release the pushbutton, and the blue LED will turn back off. Congratulations!

NOTE: The Netduino's pushbutton uses a special wiring configuration to enable it to act as both a reset button and a digital input. It also technically sends a low voltage when pushed even though your code will see a value of **true**. These values are reversed for the pushbutton inside the Netduino firmware; they are logical values instead of physical values.

Early versions of the Netduino firmware did not reverse the physical values. If your Netduino's LED exhibits the reverse behavior when you run this sample, you can either update your firmware or change the code from:

```
buttonState = button.Read()
```

to:

```
buttonState = !button.Read()
```

The exclamation mark before the word **button** reverses the result.

Conclusion

You've now created, deployed, and run your first Netduino projects. You can unplug the Netduino from your computer and demonstrate your first projects to others.

4/Expansion Shields and Electronic Components

Netduino projects consist of up to three types of components—actuators, sensors, and microchips:

- Actuators are things that do things (act). Some examples of actuators are LEDs that illuminate, motors that spin, and relay circuits that turn things on and off.

- Sensors are things that measure (sense). Some examples of sensors are light sensors (photocells), temperature sensors, and pushbuttons.

- Microchips (also known as integrated circuits or ICs) are tiny computer chips that have specialized functions. Some examples of microchips are shift registers (which allow you to hook up even more actuators and sensors), wireless networking chips, GPS chips, and digital temperature sensors.

There are hundreds of thousands of components that can be connected to a Netduino. You can find them online or at an electronics shop. You can even scavenge them from discarded electronics.

But sometimes it is nice to have these components pre-assembled so that you can focus on the logic of your app rather than the details of the components. For this reason, there are a number of expansion shields available for Netduino.

There are hundreds of shields available for Netduino. Some shields provide the circuitry needed to power motors. Some shields provide wireless networking, GPS location data, or sound capabilities. Other shields are *project shields*, which contain all the components needed to build your own robot, alcohol sensor, video game console, or other specialized project.

NOTE: There is a large community of artists and microcontroller hackers who use a platform called Arduino. Netduino has been specifically designed to be electrically compatible with both Netduino shields and most Arduino shields. If in doubt, just ask the shield vendor for sample Netduino code.

Some Netduino users build projects by using pre-assembled shields, by plugging electronic components directly into a Netduino, or by connecting a Netduino and components to a breadboard using pre-cut wires. Many Netduino users pick up a soldering iron and assemble shields and projects with wire and solder. Both are fairly easy to do and many users do both.

NOTE: If you've never soldered before, try it. It may take a few attempts to be proud of your solder joints, but it's not hard and there are plenty of videos and resources online (such as MAKE's Skill Set: Soldering; *http://blog.makezine.com/2011/01/06/skill-set-soldering/*) that make it easy to learn how to solder well.

A Gallery of Shields

Here are examples of a few expansion shields. Some come pre-assembled. Some require assembly (soldering). Also, check out *http://shieldlist.org* for a large list of Arduino-compatible shields, most of which are electrically compatible with Netduino.

Motors

Several digital pins on the Netduino can generate signals to specify the speed of motors (or the position or an arm moved by a servo). Digital pins can only provide the speed or position signal though; they don't drive enough current to power the actual motor.

A motor shield takes care of this for you by either amplifying the digital signal with another power source or by providing external power to the motor while letting the Netduino provide a separate digital control signal.

The Adafruit MotorShield, shown in Figure 4-1, requires some assembly (soldering). It's available from the Maker Shed at *http://www.makershed .com/product_p/mkad7.htm*. Motor shields are also available pre-assembled from other vendors.

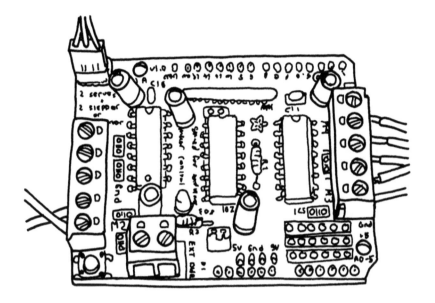

Figure 4-1. *Adafruit MotorShield*

 NOTE: Many servo motors can be operated without a motor shield. That's because they have built-in circuitry to convert digital signals into angular position (standard servos) or rotational speed (continuous rotation servos). You can connect the Netduino directly to the servo's control wire, and provide power separately to the servo. You'll learn all about servos in Chapter 7.

GPS

When building a project with wheels or when logging data while you're in motion, knowing the current location can be very useful. For these scenarios, you can use a GPS shield (Figure 4-2, available from *http://www.adafruit .com/products/98*) to provide a stream of accurate location data. Netduino community members have built both simple and sophisticated GPS parser

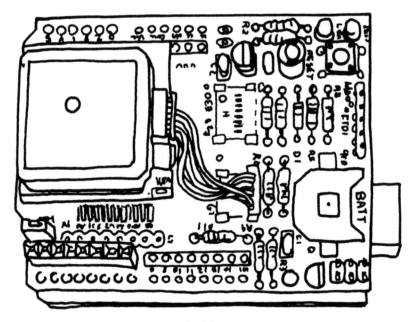

Figure 4-2. *Adafruit GPS logger shield*

code so that you can just focus on the location updates instead of worrying about the particulars of GPS's NMEA data format. For an example, check out Bob Cravens's GPS Using the Netduino post at *http://blog.bobcravens .com/2010/09/gps-using-the-netduino/*.

Wireless Networks

It's often useful to connect your Netduino to a wireless network. For those times, pick up a WiFi-compatible shield or a cellular shield (such as Seeed Studio's, shown in Figure 4-3). Sample code for these is available in the Net-duino community (for example, see *http://forums.netduino.com/index.php ?/topic/3247-seeduino-gprs-gsm-shield/*).

Graphical Display

For projects that generate lots of data and deserve a rich graphical display, there are also display shields for Netduino. Figure 4-4 shows a display shield from 4D Systems (*http://www.4dsystems.com.au/*).

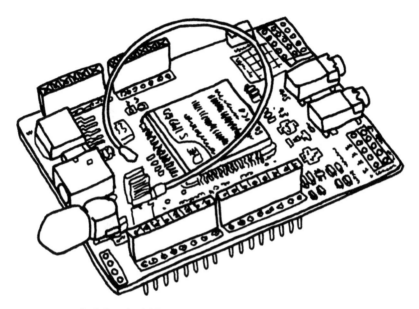

Figure 4-3. *Cellular shield from Seeed Studio*

Breadboards and Components

Another way to connect components to your Netduino is by using a breadboard or by plugging components directly into the Netduino's digital and analog pins.

If you're new to electronics, the first thing you need to know is that a breadboard, shown in Figure 4-5, has nothing to do with cutting bread. Rather, a breadboard is a nifty piece of plastic with columns and rows of holes in it. Internally, each hole is electrically connected to all the other holes in its row. This way when you push wires or legs of components into these holes, they're automatically connected to their adjacent partners.

NOTE: OK, maybe a breadboard has a little to do with cutting bread. The breadboard gets its name from the wooden breadboards that, at one time, radio makers would nail components and connectors to.

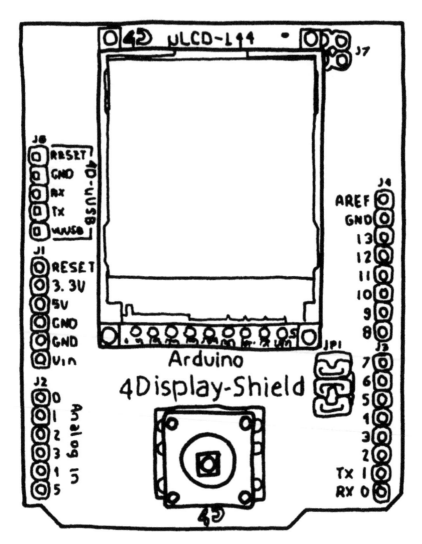

Figure 4-4. *4D Systems 1.44" display shield*

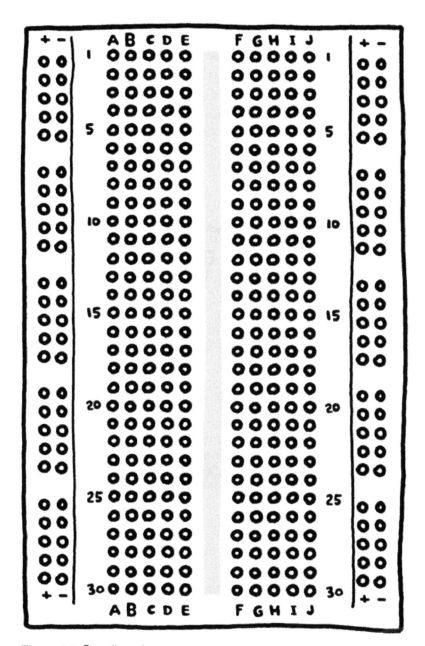

Figure 4-5. *Breadboard*

Breadboards provide a good way to quickly prototype circuits without needing to solder or make circuit boards. If you decide to make your prototyped project more permanent, you can easily pick up a prototyping PCB (a kind of "permanent breadboard") and solder each component into it instead.

If you're just plugging a component or two into your Netduino to test it, you can poke that component's leads directly into the digital or analog pin header holes on the Netduino itself.

You can use a number of electronic components in your projects, such as switches, LEDs, and motors. And often these components need a little help from other components like resistors and capacitors (see Figure 4-6).

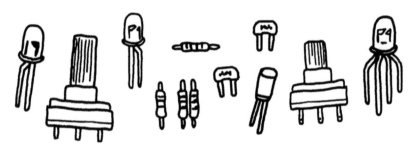

Figure 4-6. *Capacitors, resistors, and more*

Conclusion

You have now learned about expansion shields and add-on components. In the next chapter, you will interface components with your Netduino and you'll use the knob (potentiometer) on a MakerShield—a shield that's made with the Netduino in mind—to learn about measuring analog inputs.

5/Digital and Analog IO with the MakerShield

Let's take a quick look at the MakerShield, shown in Figure 5-1. When you're prototyping a project for your Netduino, a prototyping shield (such as the MakerShield) can be very useful. This shield provides a prototyping area (a good place to put a tiny breadboard) and also has built-in LEDs, a pushbutton, and a potentiometer (think "volume knob"). The MakerShield is available from Maker Shed at *http://www.makershed.com/MakerShield_p/msms01.htm*.

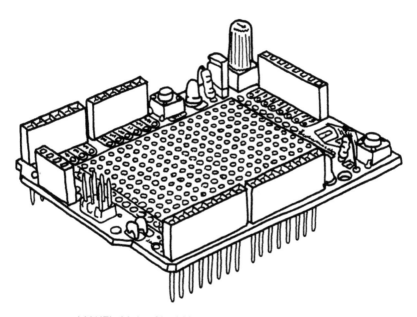

Figure 5-1. *MAKE's MakerShield kit*

NOTE: If you don't have a MakerShield, don't worry: you can plug a pushbutton and LED into a breadboard and then connect the breadboard to the Netduino using a few wires. See Chapter 6 for a complete explanation of breadboards.

To use the MakerShield, simply follow its instructions to solder it together, make sure its jumper is set in the 3.3V position, and then plug it into the top of your Netduino. Its 28 connector legs should be fairly strong—but you should always be a little gentle when inserting them.

In the bottom-left corner of the MakerShield, you'll notice a 4-pin header with the labels BTN1, LED1, LED2, and POT1 next to it. These header holes are connected to the pushbutton BTN1, to the two LEDs, and to the knob (POT = potentiometer). You'll use a short piece of solid jumper (or hookup) wire to connect these header holes to the desired digital or analog pin on your Netduino. You can get hookup wire from many places; Maker Shed has a set of deluxe breadboard wires at *http://www.makershed.com/product_p/ mkseeed3.htm*.

Pushing the MakerShield's Button

In Chapter 3, you created a project that turned on and off an LED every time a button was pushed. In that project, you used the blue LED and the pushbutton on the Netduino itself. Now that you've learned about expansion shields, breadboards, and components in Chapter 4, it's time to use them to connect an external LED and external button to your Netduino. You can also use this exact same code to work with external components.

The original code looked like this:

```
public static void Main()
{
    // write your code here
    OutputPort led = new OutputPort(Pins.ONBOARD_LED, false);
    InputPort button =
        new InputPort(Pins.ONBOARD_SW1, false, Port.ResistorMode.Disabled);

    bool buttonState = false;

    while (true)
    {
        buttonState = button.Read();
        led.Write(buttonState);
```

```
        }
    }
```

NOTE: Before you modify the project from "Pushing the Onboard Button" on page 21, you might want to make a copy of it. Visual Studio saves your projects in your Documents folder under a folder with a name like *Visual Studio 2010\Projects*. To make a copy of a project, first exit out of Visual Studio, then select the project's folder, press Control-C, and then press Control-V. If you do this with a project named *PushButton1*, you'll end up with a backup copy of it called *PushButton1 - Copy*. This doesn't actually change the project name that appears in Visual Studio's Solution Explorer, though; both copies will have the same name there.

To work with the external LED and pushbutton, you need to change a few lines of code (changes are shown in bold). Open the project again in Visual Studio, then change the line of code that sets up the LED:

```
OutputPort led = new OutputPort(Pins.GPIO_PIN_D0, false);
```

Next, change the lines of code that sets up the pushbutton:

```
InputPort button =
    new InputPort(Pins.GPIO_PIN_D1, false, Port.ResistorMode.PullUp);
```

NOTE: GPIO means *General Purpose Input/Output* port. GPIO means that a pin can be used to drive output voltage (3.3V for high, 0V for low) or can be used to interpret the input voltage (3.3V for high, 0V for low).

You changed the LED to pin "digital 0" and changed the pushbutton to pin "digital 1." That's fairly straightforward. But why did I have you change the `ResistorMode` for the pushbutton?

Pushbuttons can be wired many different ways. In one configuration, they drive the voltage high when pushed. In another popular configuration (used on the MakerShield), they drive the voltage low when pushed. Since a digital input is neither high nor low by default, you need to provide a default of high so that the MakerShield's button can drive the signal low when pushed.

By specifying `ResistorMode.PullUp`, you instruct the Netduino's microcontroller to put a tiny bit current on the pin, which makes it default to high; when the pushbutton is pressed, it overrides this signal with a strong connection to ground (0V).

Since the pushbutton will give you a reading of false (0V) when it's pressed, you need to reverse the logic in your code so that the LED lights up when the button is pressed and not vice versa. To do this, you must modify one more line of code, which you'll find inside the `while` loop:

```
buttonState = !button.Read();
```

The difference in code may seem insignificant and is easy to miss. I had you add an exclamation mark directly in front of the word `button`. In C# code, an exclamation mark means "not" or "the opposite of" a true/false (Boolean) value. If `button.Read()` returns false (because it is pushed), you want to invert that so that the button state reads true (indicating that it was pushed).

The final thing you need to do is wire up the pushbutton and LED. To do this, connect a short piece of wire between digital header pin D0 on the Maker-Shield and the header hole labeled LED1 or LED2 (either one is fine). Also, connect a short piece of wire between digital header pin D1 on the Maker-Shield and the header hole labeled BTN1. Figure 5-2 shows how to hook things up.

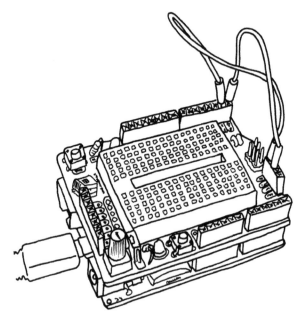

Figure 5-2. *MakerShield's button and LED wired up*

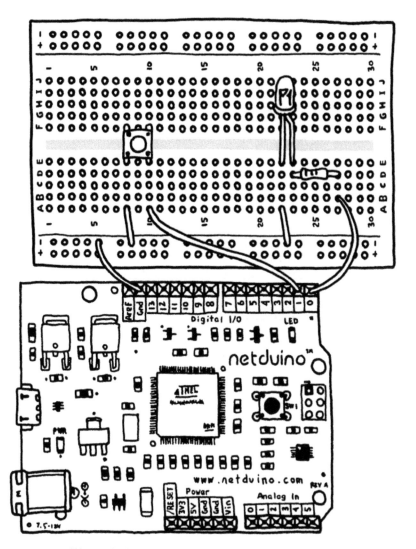

Figure 5-3. *Wiring the button and LED up on a breadboard*

 NOTE: If you don't have a MakerShield, you can still lay out this project on a breadboard. Figure 5-3 shows how you'd lay it out on a breadboard (for the resistor, you can use one with a value between 150 and 330 ohm).

Now run your project as you did in Chapter 3. Press the Start Debugging button in the toolbar at the top of the screen (or press F5).

When you press the pushbutton on your MakerShield, the LED will turn on; release the pushbutton and it will turn off.

Analog Inputs

You have learned how to sense the value of a digital input (with a state of true or false). That works great for pushbuttons and open/close sensors, but the world is full of in-between values.

To measure real-world values like temperature, humidity, light levels, or the position of a knob, you need an *analog input*. With Netduino, analog sensors output a voltage from 0V to 3.3V. The Netduino's analog inputs translate these voltages into a range of values (typically 0 through 1023). Using analog inputs, you can discern a temperature value in degrees, the volume you should play music at, or even particular shades of color.

Measuring Voltage

The MakerShield has an integrated potentiometer (like a volume knob). A potentiometer has three pins: one for an input voltage (such as 3.3V), another for ground (0V), and one that is connected to the *wiper* inside the potentiometer. You connect one of the Netduino's analog pins to the wiper. When the potentiometer's knob is turned, the output voltage varies (from 0V at one side, 1.65V in the middle, and 3.3V turned all the way). The changes are linear; as a result, you can measure the position of the knob based on its output voltage.

You can read this voltage in your Netduino app and change program behavior based on its value. In this section, you'll build an app to read the analog values, and then you'll change the blinking speed of one of the MakerShield's LEDs to follow the potentiometer's position.

First, write the code. Create a new project as described in Chapter 3 and open the *Program.cs* file in the main code editing window.

Now, type the following right after the line that reads `// write your code here`:

```
OutputPort led = new OutputPort(Pins.GPIO_PIN_D0, false);❶
AnalogInput pot = new AnalogInput(Pins.GPIO_PIN_A0);❷

int potValue = 0;❸
```

❶ As before, the first line creates an `OutputPort` so you can turn on and off one of the LEDs on the MakerShield.

❷ The second line creates an `AnalogInput` named `pot` (the standard nick-name for a potentiometer). It specifies analog pin 0 (`GPIO_PIN_A0`), which you'll wire up to the potentiometer shortly.

❸ The third line of code creates a variable named `potValue`. The word *int* that precedes the variable indicates that `potValue` will store an integer value. Integer values are 32 bits and can store values between roughly negative 2 billion and positive 2 billion. You'll only use numbers 0 through 1023 here.

Next, you need to read the potentiometer's value and then vary the LED's blinking speed based on that value. You'll do this in an infinite loop as before. Add the following to your program:

```
while (true)
{
    // read the value of the potentiometer
    potValue = pot.Read();

    // blink the led based on the potentiometer's value (0-1023ms)
    led.Write(true);
    Thread.Sleep(potValue);
    led.Write(false);
    Thread.Sleep(potValue);
}
```

Next, connect the LED and potentiometer to the proper pins on the Netduino. Plug your MakerShield on top of your Netduino if it's not already. Connect a short piece of wire between digital header pin D0 on the MakerShield and the header hole labeled LED1 or LED2 (either one is fine). Also, connect a short piece of wire between analog header pin A0 on the MakerShield and the header hole labeled POT1. Figure 5-4 shows the connections. If you don't have a MakerShield, you can lay this circuit out on a breadboard as shown in Figure 5-5.

Now run your project. Turn the potentiometer's knob to the left, to the center, and all the way to the right. Watch how the speed of the LED's blinking pattern changes as you adjust the knob.

Before you move onto another analog sensor, try one more thing. As you may have noticed, the analog input provides a range of 0-1023 by default. But perhaps you want a different range that represents something in the real world (like temperature). For this experiment, let's change the potentiometer value's range to 100-250. This will create a more pleasing blinking range.

Figure 5-4. *MakerShield's potentiometer and LED wired up*

In your main routine, immediately after creating the `AnalogInput` pot, add the following line of code:

```
pot.SetRange(100, 250);
```

Your final code should look like this:

```
public static void Main()
{
    // write your code here
    OutputPort led = new OutputPort(Pins.GPIO_PIN_D0, false);
    AnalogInput pot = new AnalogInput(Pins.GPIO_PIN_A0);
    pot.SetRange(100, 250);

    int potValue = 0;

    while (true)
    {
        // read the value of the potentiometer
        potValue = pot.Read();

        // blink the led based on the potentiometer's value (0-1023ms)
        led.Write(true);
        Thread.Sleep(potValue);
        led.Write(false);
        Thread.Sleep(potValue);
```

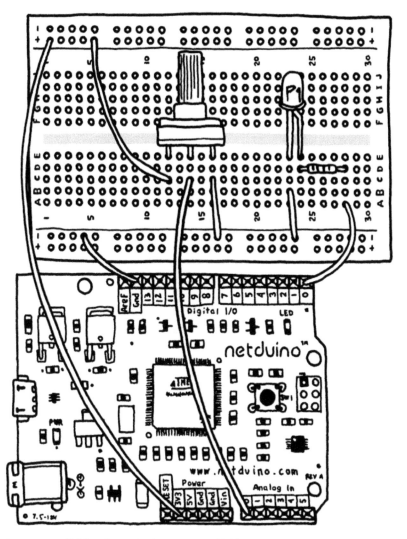

Figure 5-5. *Wiring the potentiometer and LED up on a breadboard*

        }

    }

Now run your project again. You'll notice that the blinking goes from fast to slower but the changes are much more gradual. To further experiment, reverse the values in the SetRange command, and you'll have effectively flipped the potentiometer in reverse!

Other Analog Sensors

In addition to analog sensors that measure a physical position like potenti-
ometers, there are analog sensors that measure real-world conditions like
photocells (for light level), force sensors, and even color sensors.

The nice thing about analog sensors is that they all work the same way: plug
their output into one of the Netduino's analog input pins, set the desired
range, and read the sensor's value.

NOTE: The Netduino's analog input ports are designed to handle voltages between 0V and 3.3V.
Some analog sensors output higher voltages. You
should always check the specifications for an analog sensor before plugging it into your Netduino. If
unsure, you can use a multimeter to measure the
output voltage.

Conclusion

You have now interfaced external components with your Netduino and successfully learned how to use a expansion shield. In the next chapter, you'll
learn how to do a few more things with LEDs.

6/Breadboards and LEDs

Like many light bulbs, LEDs can not only be turned on and off; they can be dimmed (or made to look dimmed) as well.

By changing the intensity of LEDs, you can represent the value of a sensor, create cool effects like pulsing and breathing, or signal that your Netduino app has important information to share.

You can also group LEDs together to represent numbers. And you can combine colored LEDs together to form new colors. You can even dim those colored LEDs, generating millions of color combinations.

Changing Intensity

To dim an LED (i.e., change its intensity), you simply turn it on and off quickly. If the LED is on half the time and off half the time, it will appear about half as bright to the human eye. If you turn it on only 10% of the time, it will seem very dim. While it is possible to simply turn LEDs on and off rapidly using looping on and off commands in code, the Netduino's microcontroller has an integrated feature especially suited for this task. This feature is called *Pulse Width Modulation* (PWM) and, if you tell it the percentage of time to toggle the LED on, it will automatically dim the LED by pulsing it rapidly. This is all done in the background, which frees up both the microcontroller and your code to focus on more important tasks.

Previously, you turned on and off an LED using the OutputPort feature of .NET Micro Framework. Now you're ready to update that code to set the LED's intensity using the PWM feature instead.

First, you'll need to wire up an LED and a potentiometer using a breadboard.

 NOTE: If you have a MakerShield, you can use its integrated potentiometer and one of its LEDs for this example. Chapter 5 tells you everything you need to know about the MakerShield.

Setting Up the Breadboard

First, connect the 3.3V header from your Netduino to the voltage rail column of your breadboard. The voltage rail is on the side of the breadboard and is typically designated by a red line running alongside it. If you've rotated the breadboard as in Figure 6-1, which shows the completed connections, the rails will be on the top and bottom.

If you don't have a rail that's marked red, or if your breadboard has only one rail on each side, pick one rail to use as the voltage rail. To do this, plug a wire into the 3.3V header on your Netduino and then into any of the header holes in the breadboard's voltage rail column.

When you provide power to a rail on the side of a breadboard, all of the holes in that column will become power supplies. Columns on the sides of breadboards are internally connected to provide this feature.

 NOTE: Some large breadboards add multiple power rails. If this is the case with your breadboard, you may need to connect rail segments with wire to extend your power source.

Now that you've supplied 3.3V power to the breadboard's power rail, you need to provide a return path for the electricity as well. Connect one of the GND (ground) headers from your Netduino to one of the header holes in the ground rail column of your breadboard. The ground rail is typically next to the power rail and designated by a blue (sometimes black) line, but if your breadboard has only one rail on each side, use the bottom rail as the ground rail.

Now that you have power on your breadboard, you can hook up the potentiometer and LEDs.

Hooking Up the Components

Push the pins of a potentiometer into one side of the breadboard.

Since each breadboard row is connected internally, you'll want to make sure that no two pins from your potentiometer are plugged into the same row. If your potentiometer's pins are plugged into the same row, simply remove it and rotate it 90 degrees.

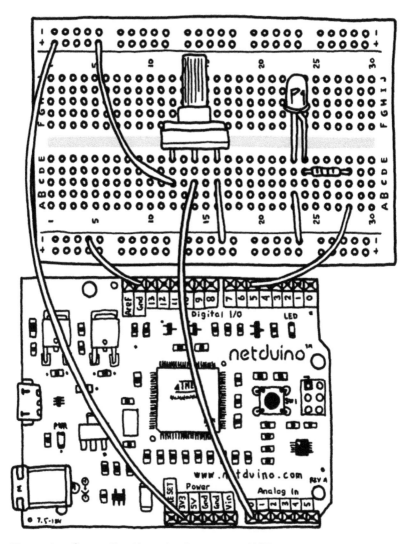

Figure 6-1. *Connecting the potentiometer and LED*

Please note that breadboard rows are often split into two sides separated by an empty middle section; these sides are independent and the rows on each side are not connected to each other.

There are three pins on a potentiometer: power, output, and ground.

NOTE: The power and ground pins on a potentiometer and often interchangeable: if you plug them in backward, your output values will generally be reversed to match. If in doubt as to which pins to use, check the potentiometer's datasheet or instructions.

Connect a wire from the breadboard's 3.3V power rail to the first pin (power pin) of your potentiometer. Plug one end of the wire into any header hole in the power rail and plug the other end into any hole in the same row as your potentiometer's first pin (power pin). Since each row in the breadboard is electrically connected internally, you are connecting the power rail to the pin even though they are not touching.

Similarly, connect another wire between the breadboard's ground rail and the third pin (ground pin) of your potentiometer. Finally, connect a longer wire from the potentiometer's output (center pin) to analog pin A0 on your Netduino.

NOTE: If you're using a MakerShield, you can connect a wire between the A0 and POT headers to hook up the built-in potentiometer instead.

Now you'll plug in an LED. Every LED has a longer positive leg (anode) and a shorter negative leg (cathode). It is important to wire these up correctly, as LEDs do not work when reversed and can be damaged if connected improperly.

Plug the LED's two legs into a breadboard. Next, connect a wire between a digital output on the Netduino and the positive (longer) leg on the LED. For this example, you need to use digital pin D5.

NOTE: If you're using a MakerShield, you can connect a wire between the D5 and LED1 header to hook up a built-in LED instead.

Then, you need to connect the LED's negative (shorter) leg to the ground rail, but not directly. Although you could technically connect it directly, and everything will appear to work fine, it could accidentally let us drive more power through the LED than it is rated for and could damage the LED. So instead of connecting the LED's negative leg directly to the ground rail with a wire, connect it with a resistor instead. For this example, use a 150 ohm resistor (if you don't have a 150 ohm resistor, anything between 150 and 330 ohms will do).

 NOTE: Assuming you're using an LED with a 2.1 volt *forward voltage drop*, a 150 ohm resistor will limit the current to 8 *milliamperes* (mA). 8mA is the maximum current rating of most of the Netduino's digital pins, so this is a perfect value.

Plug one end of the resistor into the same breadboard row as the LED's negative leg and then plug the other end of the resistor into the ground rail. Resistor legs are interchangeable, so don't worry about which leg gets plugged into the ground rail. Figure 6-1 shows the circuit.

 NOTE: Since you're using the Netduino's digital pin D5, which is rated for more current, you could use an 82 ohm resistor and roughly double the current to the LED. For more information on how to calculate the best resistor values for LEDs, see *http://ledcalculator.net/*.

Writing the Dimmer Code

Create a new project as before and open the *Program.cs* file in the main code editing window.

Now, type the following:

```
PWM led = new PWM(Pins.GPIO_PIN_D5);❶
AnalogInput pot = new AnalogInput(Pins.GPIO_PIN_A0);❷
pot.SetRange(0, 100);
int potValue = 0;❸
```

❶ This line of code creates a PWM object using our digital pin D5. This lets you control the intensity of the LED by changing its duty cycle (percentage of time turned on vs. off).

❷ The second and third lines of code set up the potentiometer as before, but this time with a range of 0 through 100. This will give you a percentage that you can use to set the LED's intensity.

❸ This line creates a variable named `potValue`, which you'll use to store and retrieve the potentiometer's current position.

Now you need to write some code to change the LED's intensity based on the value of the potentiometer. Add the following:

```
while (true)
{
    // read the value of the potentiometer
    potValue = pot.Read();

    // change the led intensity based on
    // the potentiometer's value (0-100%)
    led.SetDutyCycle((uint) potValue);
}
```

Your final code will look like this:

```
public static void Main()
{
    // write your code here
    PWM led = new PWM(Pins.GPIO_PIN_D5);
    AnalogInput pot = new AnalogInput(Pins.GPIO_PIN_A0);
    pot.SetRange(0, 100);
    int potValue = 0;

    while (true)
    {
        // read the value of the potentiometer
        potValue = pot.Read();

        // change the led intensity based on
        // the potentiometer's value (0-100%)
        led.SetDutyCycle((uint) potValue);
    }
}
```

Now run the project on your Netduino (don't forget to set the deployment target using the project's properties dialog first). Turn the potentiometer's knob to the left, to the center, and all the way to the right. Watch how the brightness of the LED changes as you adjust the knob.

NOTE: Digital pins D5, D6, D9, and D10 can all be used as PWM-enabled pins.

Mixing Colors

Every dot on a television, computer monitor, or phone screen can display one of many thousands (or millions!) of colors. This is generally done by mixing tiny red, green, and blue pixels together. By blending various intensities of each of these three basic colors, you can display a whole spectrum of colors. Let's do so now.

For this example, you'll need an RGB LED. This is an LED with red, green, and blue LED elements inside of it. It will generally have four legs: one positive input leg for each of the three colors and one negative output leg for ground. The negative (longest) leg is called the *common cathode*. There are also *common anode* versions of RGB LEDs that you can use with Netduino, but they need to be wired in reverse and use inverse programming logic.

Since each colored LED element needs its own resistors, you'll wire up resistors between the microcontroller and the RGB LED on this example.

 NOTE: Resistors can be placed either before or after the LED in a circuit. They will limit the amount of power driven through the resistors either way. But you must make sure to include them, or you risk damaging the LED. In the case of RGB LEDs, the resistors must be added to the anode side as we have done here.

Plug an RGB LED into a breadboard. Make sure all four legs are plugged into different rows. Connect a wire between the negative (longest) leg and the ground rail.

Now, connect a resistor between each of the three RGB positive legs and a separate row on the other side of the breadboard's middle divider. Finally, connect a wire between each of those rows and a digital pin on the Netduino. Use digital pins D5, D6, and D9 on the Netduino.

Now you're ready to write some code to display the color purple. Start out by typing the following:

```
PWM redLed   = new PWM(Pins.GPIO_PIN_D9);
PWM greenLed = new PWM(Pins.GPIO_PIN_D6);
PWM blueLed  = new PWM(Pins.GPIO_PIN_D5);
```

These three lines of code will set up the **redLed**, **greenLed**, and **blueLed** objects. Those PWM objects will let you control the intensity of each of our three LEDs.

Don't worry if you're not sure if the red, green, and blue LEDs are plugged into the correct digital pins. If they're arranged in the wrong order, you can swap them around in a few moments. However, many RGB LEDs are laid out in the way shown in Figure 6-2: R, common cathode, G, B.

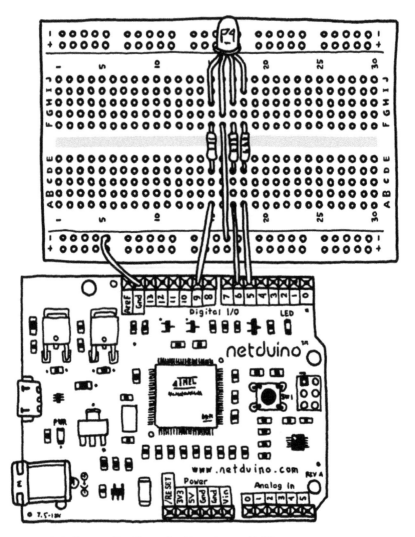

Figure 6-2. *Connecting the potentiometer and LED*

Continue by setting the colors to the intensities for the color purple:

```
// change the color intensities
redLed.SetDutyCycle(60);    // 60% red intensity
greenLed.SetDutyCycle(0);   // 0% green intensity
blueLed.SetDutyCycle(100);  // 100% blue intensity
```

Finally, add a line of code that puts the program to sleep indefinitely. This makes it so that your program doesn't just stop running.

```
// go to sleep
Thread.Sleep(Timeout.Infinite);
```

Your final code will look like this:

```
public static void Main()
{
    // write your code here
    PWM redLed   = new PWM(Pins.GPIO_PIN_D9);
    PWM greenLed = new PWM(Pins.GPIO_PIN_D6);
    PWM blueLed  = new PWM(Pins.GPIO_PIN_D5);

    // change the color intensities
    redLed.SetDutyCycle(60);    // 60% red intensity
    greenLed.SetDutyCycle(0);   // 0% green intensity
    blueLed.SetDutyCycle(100);  // 100% blue intensity

    // go to sleep
    Thread.Sleep(Timeout.Infinite);
}
```

Now run the program. If you see the color purple, congratulations! Stop the program, change the intensity values to different percentages, and run it again. You'll discover a whole world of colors.

If the color is incorrect, stop the program. Change the intensity of one color to 100 and the other two colors to 0. Rerun the program. If the wrong LED lights up, swap the three wires between D5/D6/D9 until the correct colors light up. Then repeat for the next color(s). Your colors should now be wired up correctly.

 NOTE: You can look up the red, green, and blue intensity values for colors with an online color calculator. One such resource is *http://www.rapidtables.com/web/color/RGB_Color.htm*.

Conclusion

You've now gotten some more experience hooking up components to your Netduino, and know your way around a breadboard. You've also gotten to play around with some different ways of creating light with LEDs, one of the most popular actuators used with microcontrollers. In the next chapter, you'll learn about some other actuators: those that create sound and motion.

7/Sound and Motion

Speakers generate sounds by vibrating.

You can generate those sound vibrations with the same PWM feature you used in Chapter 6 to adjust the intensity of LEDs. By changing the period and length of pulses, you can change the frequency of the sounds. And by using pulses that correspond to musical notes, you can play songs.

You can also use the length of pulses as a signal to a servo (motor), setting the servo to certain angular positions through specific pulse lengths.

Making Music

To play a song, you'll need a speaker. For this example, use a common and inexpensive piezo speaker.

Wire up the piezo's positive wire to pin D5 on your Netduino. Wire up the piezo's negative (ground) wire to the GND pin on your Netduino. Figure 7-1 shows the connections.

 NOTE: Some piezos have legs rather than wires. In most cases, one leg is longer than the other, and the long leg is positive. There may also be faint markings on the piezo indicating which is positive and which is negative. If in doubt, check your piezo's instructions or datasheet for proper wiring.

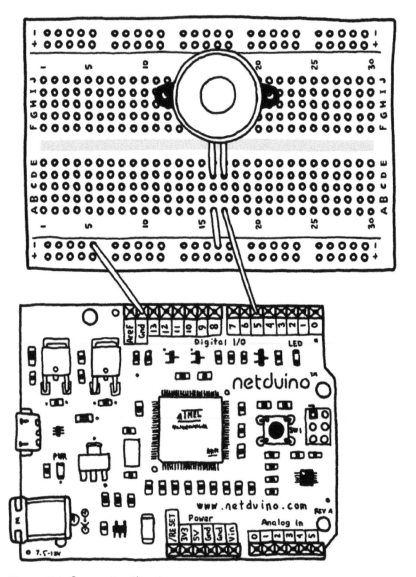

Figure 7-1. *Connecting the piezo*

Create a new Netduino project. Add the following code to the top of your
Main routine:

```
// store the notes on the music scale and their associated pulse lengths
System.Collections.Hashtable scale = new System.Collections.Hashtable();❶
```

```
// low octave
scale.Add("c", 1915u);❷
scale.Add("d", 1700u);
scale.Add("e", 1519u);
scale.Add("f", 1432u);
scale.Add("g", 1275u);
scale.Add("a", 1136u);
scale.Add("b", 1014u);

// high octave
scale.Add("C", 956u);
scale.Add("D", 851u);
scale.Add("E", 758u);

// silence ("hold note")
scale.Add("h", 0u);
```

❶ I'm introducing a new concept here called a Hashtable. A Hashtable is a collection of keys and values. The keys here are the notes, and the numbers are the pulse length values. Hashtables make it easy to look up values by specifying their keys. You'll do this later by looking up notes and getting back their pulse length values.

❷ The "u" after each value signifies that the values should be stored as unsigned integers (whole numbers). You could store this data in other ways, but this makes the sample simpler to understand later on.

NOTE: These note values are calculated by the equation *1 / (2 * toneFrequency)*.
toneFrequency is the number of cycles per second for a specific note. You can extend this scale further if desired.

Next, specify the speed of the song. You'll specify this in beats per minute, a common measurement for music.

```
int beatsPerMinute = 90;
int beatTimeInMilliseconds =
  60000 / beatsPerMinute; // 60,000 milliseconds per minute
int pauseTimeInMilliseconds = (int)(beatTimeInMilliseconds * 0.1);
```

To calculate the beatTime (length of a beat) in milliseconds, divide 60,000 (the number of milliseconds in a minute) by the number of beats per minute.

And since you want a tiny bit of pause in between each note (so that the notes don't all blur together), multiply the beat length by 10% to create a pause length. Later on, you'll subtract 10% from the beat length and insert this pause instead, keeping a steady rhythm but making the song more pleasant.

NOTE: You *casted* the pause time to an integer value. Integers are positive or negative whole numbers. Since 0.1 (10%) is a fractional number and not a whole number, you need to do this explicitly. The programming language is smart and realizes that you might lose a little bit of accuracy by ending up with a few tenths of a number that can't be stored in an integer. So it requires that you explicitly *cast* the product into an integer to show you know that you might lose some precision.

Now, define the song. This song involves animals on a farm and the sounds that they make:

```
// define the song (letter of note followed by length of note)
string song = "C1C1C1g1a1a1g2E1E1D1D1C2";
```

The song is defined as a series of notes (upper-case for higher octave, lower-case for lower octave) followed by the number of beats for each note. For the musicians' reference, one beat is a quarter note in this instance and two beats is a half note, in 4/4 time.

Now that you have a musical scale and a song, you need a speaker to play them. You'll add one here:

```
// define the speaker
PWM speaker = new PWM(Pins.GPIO_PIN_D5);
```

You're all set to play the song. Create a loop that reads in each of the note/length entries in the song string, and then plays them by generating specific pulses for the piezo using PWM.

Create a loop:

```
// interpret and play the song
for (int i = 0;❶ i < song.Length;❷ i += 2❸)
{
    // song loop

}
```

❶ The first part of the for statement, `int i = 0`, establishes the variable `i` as a position counter in the encoded song's string.

❷ The middle part of the statement, `i < song.Length`, repeats the loop until you've finished playing the song.

❸ The third part of the statement, `i += 2`, moves the position forward by two places (the size of a note/length pair) each time through the loop.

Inside this loop, you'll extract the musical notes and then play them. Inside the pair of curly braces, add the music interpreter/player code:

```
// extract each note and its length in beats
string note = song.Substring(i, 1);❶
int beatCount = int.Parse(song.Substring(i + 1, 1));❷
```

❶ The first line reads the note out of the song string at the current position, i. Song.Subtring(i, 1) means "the string data starting at position i, length of 1 character."

❷ The second line reads the beat count of the song string. It reads the beat count at one position beyond the note. Then, it uses int.Parse() to translate this letter into an integer number that you can use. This is needed because you can't do math with letters, and you'll need to do a bit of math with the beat count shortly.

Now that you have the note, look up its duration:

```
// look up the note duration (in microseconds)
uint noteDuration = (uint)scale[note];
```

By passing in the letter (the variable note), you get back the noteDuration. Lookup is a powerful feature of C# that can be used for many purposes.

Whenever you retrieve a value from a collection (a Hashtable in this instance), you need to explicitly cast it since the programming language wants to be sure it knows what it's getting.

Now, play the note:

```
// play the note for the desired number of beats
speaker.SetPulse(noteDuration * 2, noteDuration);
Thread.Sleep(
  beatTimeInMilliseconds * beatCount - pauseTimeInMilliseconds);
```

SetPulse is another PWM feature (in addition to SetDutyCycle, which you saw in Chapter 6). It sets a period of twice the noteDuration, followed by a duration of noteDuration. This will generate a pulse at the frequency required to sound the desired note.

You use Thread.Sleep to keep the note playing for the specified number of beats. You subtracted pauseTimeInMilliseconds from the note's sounding time so that you can pause a tenth of a beat and make the music more pleasant.

Now that you've played the note, go ahead and insert that one-tenth beat pause:

```
// pause for 1/10th of a beat in between every note.
speaker.SetDutyCycle(0);
Thread.Sleep(pauseTimeInMilliseconds);
```

By setting the duty cycle to zero, you turn the piezo off momentarily. This creates the nice pause.

Finally, skip past the curly brace that ends the for loop and add one more line of code:

```
Thread.Sleep(Timeout.Infinite);
```

This will pause the program after playing the tune.

The final code should look like this:

```
public static void Main()
{
    // write your code here

    // store the notes on the music scale and their associated pulse lengths
    System.Collections.Hashtable scale = new System.Collections.Hashtable();

    // low octave
    scale.Add("c", 1915u);
    scale.Add("d", 1700u);
    scale.Add("e", 1519u);
    scale.Add("f", 1432u);
    scale.Add("g", 1275u);
    scale.Add("a", 1136u);
    scale.Add("b", 1014u);

    // high octave
    scale.Add("C", 956u);
    scale.Add("D", 851u);
    scale.Add("E", 758u);

    // silence ("hold note")
    scale.Add("h", 0u);

    int beatsPerMinute = 90;
    int beatTimeInMilliseconds =
      60000 / beatsPerMinute; // 60,000 milliseconds per minute

    int pauseTimeInMilliseconds = (int)(beatTimeInMilliseconds * 0.1);

    // define the song (letter of note followed by length of note)
    string song = "C1C1C1g1a1a1g2E1E1D1D1C2";

    // define the speaker
    PWM speaker = new PWM(Pins.GPIO_PIN_D5);

    // interpret and play the song
    for (int i = 0; i < song.Length; i += 2)
    {
```

```
    // extract each note and its length in beats
    string note = song.Substring(i, 1);
    int beatCount = int.Parse(song.Substring(i + 1, 1));

    // look up the note duration (in microseconds)
    uint noteDuration = (uint)scale[note];

    // play the note for the desired number of beats
    speaker.SetPulse(noteDuration * 2, noteDuration);
    Thread.Sleep(
      beatTimeInMilliseconds * beatCount - pauseTimeInMilliseconds);

    // pause for 1/10th of a beat in between every note.
    speaker.SetDutyCycle(0);
    Thread.Sleep(pauseTimeInMilliseconds);
  }

  Thread.Sleep(Timeout.Infinite);
}
```

Now, run your app. Sing along with the song if you'd like. Look up the full song online and add the rest if desired.

Motors and Servos

There are several types of electromechanical devices that generate motion. Two of the most common are motors and servos.

Motors work by using the PWM DutyCycle, just like LEDs. You can use a motor shield or design your own circuitry to connect motors to your Netduino. Just like LEDs, when you increase the DutyCycle motors spin faster; when you decrease the DutyCycle, they slow down.

Circuitry to power motors generally uses an H-Bridge, which is effectively a device that uses the PWM output of your Netduino to supply a much larger amount of power at a voltage proportional to your PWM signal's "on time." The PWM signal is combined with a GPIO input specifying whether the motor should spin forward and backward. All together these create a negative or positive analog voltage that spins the motor.

The drawback of using motors is the lack of precision. You can spin them faster or slower, but often you want something to move at a specific speed or move to a specific position. When you want to move between positions, you want to use servos.

Servo Control

Servos create motion by turning forward or backward to specific positions. To demonstrate, I'll show you a program that moves through various points of a servo's position range.

To wire up the servo, connect its red wire to the 5V header on your Netduino and its GND wire (usually black, but it might be brown) to a Ground header. Then, connect the signal wire (usually white, but may be another color) to pin D5. Figure 7-2 shows the connections. You'll use PWM signals to specify the servo's position.

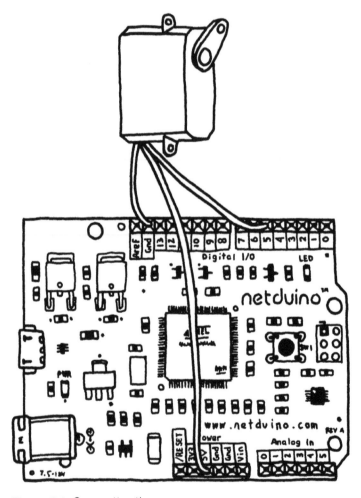

Figure 7-2. *Connecting the servo*

Create a new Netduino project. Add the following code:

```
PWM servo = new PWM(Pins.GPIO_PIN_D5);
```

As before, this creates a new PWM object using pin D5. You'll use specific pulse lengths to direct the servo to move to specific positions.

Now, set some boundaries for the servo. The following values are the number of microseconds in a pulse, which represent the lowest and highest position values for the servo. This will represent a full 90 degrees of motion:

```
uint firstPosition = 1000;
uint lastPosition = 2000;
```

 NOTE: Some servos offer extended range, such as 180 degrees. These servos may use the range of 0-1,000 to represent the extra 90 degrees of range.

Next, move through the full range of motion for the servo. Add the following code:

```
// move through the full range of positions
for (uint currentPosition = firstPosition;
     currentPosition <= lastPosition;
     currentPosition += 10)❶
{
     // move the servo to the new position.
     servo.SetPulse(20000, currentPosition);❷
     Thread.Sleep(10);
}
```

❶ The for loop is similar to the one you used for the piezo earlier in the chapter. But instead of using i as the counter variable, you're using currentPosition. It is good programming practice to give variables descriptive names like this (although i is commonly used for a counter variable, so it's okay that it wasn't as descriptive as this one).

You loop from the first position to the final position, at 10 unit intervals (approximately 0.9 degrees each).

❷ Inside the loop, you move the servo to the new position and then sleep for ten milliseconds. This will give you a smooth motion.

Now, return the servo to its first position. You can move the servo backward just as easily as forward:

```
// return to first position and wait a half second.
servo.SetPulse(20000, firstPosition);
```

Finally, finish up the app by putting it to sleep:

```
Thread.Sleep(Timeout.Infinite);
```

Your final app should look like this:

```
public static void Main()
{
    // write your code here
    PWM servo = new PWM(Pins.GPIO_PIN_D5);

    uint firstPosition = 1000;
    uint lastPosition = 2000;

    // move through the full range of positions
    for (uint currentPosition = firstPosition;
        currentPosition <= lastPosition;
        currentPosition += 10)
    {
        // move the servo to the new position.
        servo.SetPulse(20000, currentPosition);
        Thread.Sleep(10);
    }

    // return to first position and wait a half second.
    servo.SetPulse(20000, firstPosition);

    Thread.Sleep(Timeout.Infinite);
}
```

Now run your app. You can now create motion digitally!

This specific demo may remind you of the action of an old typewriter or dot matrix printer; for those not acquainted with either, they are featured in many older movies.

 NOTE: Did you notice that the power LED on the Netduino flickered during servo movement? If so, that's because you connected a servo directly to Netduino's 5V supply power header. Servos use a lot of power—especially when they meet resistance —so you'll usually want to give them a separate power supply. If you do this, you may need to connect the ground (GND) pins from both power supplies together.

Conclusion

You have now made music with a piezo and made motion with a servo. It's amazing how many electronics can be controlled through simple pulses of electricity.

Next up: connecting your Netduino Plus to the Internet.

8/Connecting to the Internet

Reading values from sensors, flashing LEDs, and moving are all very useful and cool things to do. But connecting these up to the Internet opens up a world of possibilities.

You can create tweeting sensors, phone apps for your Netduino projects, and more.

 NOTE: Connecting to the Internet is a big topic, which could take up a book of its own. And it does. Those up to speed with C# programming can learn more via O'Reilly's book *Getting Started with the Internet of Things*. This book, written by Dr. Cuno Pfister, details how to link a world of sensors and actuators together over the Internet, generate reports from that data, work with router firewalls, and more.

To introduce this topic, you'll create a Netduino app that simply turns on and off its onboard LED in response to a web page request. Yes—you can serve up web content from a Netduino!

For this sample, you'll need a Netduino Plus (which includes integrated networking).

Coding the Server

Since you're using the Netduino Plus's built-in LED for this demo, there's no need to wire up any components. This means you can move on to the code.

Create a new Netduino Plus app. Make sure that you select "Netduino Plus" instead of "Netduino," so that the networking features are available to your app.

Inside your Main routine, start by typing in the following code:

```
// set up the LED and turn it off by default
OutputPort led = new OutputPort(Pins.ONBOARD_LED, false);
```

As before, this configures the onboard LED. You'll turn this on and off via web requests.

Next, define a port for your Netduino to listen on. Port 80 is the default web server port, so use it here. If you want to connect to your Netduino over the Internet you may need to use a higher port number such as 8080—or use your router's port forwarding feature—so that you can access your Netduino through a firewall:

```
// configure the port # (the standard web server port is 80)
int port = 80;
```

For this example, insert a five-second wait. This is so that the Netduino has time to request an IP address from the local router. Experienced C# programmers may want to write some code that waits only as long as it takes to get the address—or they may want to use a static IP address instead:

```
// wait a few seconds for the Netduino Plus to get a network address.
Thread.Sleep(5000);
```

Then, display the Netduino's IP address in the Visual Studio programming environment. You need this so that you'll know which address to use to talk to the Netduino Plus from your web browser:

```
// display the IP address
Microsoft.SPOT.Net.NetworkInformation.NetworkInterface
  networkInterface =
    Microsoft.SPOT.Net.NetworkInformation.NetworkInterface.
    GetAllNetworkInterfaces()[0];

Debug.Print("my ip address: " + networkInterface.IPAddress.ToString());
```

The first line of code gets a reference to the first (and only) network interface on the Netduino Plus. The second line prints that address in the Output window of Visual Studio.

 ### Summoning the Output Window

If you don't see the Output window, press Ctrl+Alt +O. If that still doesn't work, you'll want to enable Expert mode in Visual Studio Express (Tools→Settings→Expert Settings).

Now that you know the Netduino Plus's IP address, you can have it start listening for incoming web page requests:

```
// create a socket to listen for incoming connections
Socket listenerSocket = new Socket(AddressFamily.InterNetwork,
                                   SocketType.Stream,
                                   ProtocolType.Tcp);❶
IPEndPoint listenerEndPoint = new IPEndPoint(IPAddress.Any, port);❷

// bind to the listening socket
listenerSocket.Bind(listenerEndPoint);❸
// and start listening for incoming connections
listenerSocket.Listen(1);❹
```

❶ This line of code creates a socket, `listenerSocket`, which will soon listen for incoming requests.

❷ This line specifies the endpoint; the combination of network address(es) and port that will be used.

❸ This line binds to that endpoint to prepare the Netduino for incoming messages.

❹ The final line of code listens for incoming requests. I have specified a maximum backlog of "one waiting connection," but you can increase this a little bit as you build projects of your own.

Now that the Netduino Plus is listening for incoming requests, create an infinite loop in which you watch for and then respond to those incoming requests.

Add the following loop:

```
// listen for and process incoming requests
while (true)
{

}
```

Inside this loop, you need to handle each individual request. First, wait for a connection:

```
// wait for a client to connect
Socket clientSocket = listenerSocket.Accept();
```

Then, wait up to five seconds for data to arrive:

```
// wait for data to arrive
bool dataReady = clientSocket.Poll(5000000, SelectMode.SelectRead);
```

Before you process that incoming data, make sure you received something. You will put all of this processing code inside of another subsection of code:

```
// if dataReady is true and there are bytes available to read,
// then you have a good connection.
if (dataReady && clientSocket.Available > 0)
{

}
```

As long as data is ready and there are more than zero bytes of a request available, you can process the data.

Inside the if statement's curly braces, read all the incoming data:

```
byte[] buffer = new byte[clientSocket.Available];
int bytesRead = clientSocket.Receive(buffer);
```

Those lines of code first create a buffer, which will hold the incoming request. Then you receive that data into the buffer. The **bytesRead** value will hold the number of bytes that you actually read.

 NOTE: You'll usually want to create smaller buffers to read data, rather than reading it all at once. But I know this request should be relatively small, so I'm simplifying this project by reading the full request in one go.

Now, turn the buffer into a string that you can process easily:

```
string request =
  new string(System.Text.Encoding.UTF8.GetChars(buffer));
```

The keyword that you're looking for is the text "ON" (or "OFF"), in all upper case:

```
if (request.IndexOf("ON") >= 0)
{
    led.Write(true);
}
else if (request.IndexOf("OFF") >= 0)
{
    led.Write(false);
}
```

With this code, if the user requests a web page and adds the text "ON" in the request, the LED will turn on; if the user requests a web page and adds the text "OFF," it will turn off.

Now that you've turned on/off the LED, write its status to a string that you can return to the user's browser:

```
string statusText = "LED is " + (led.Read() ? "ON" : "OFF") + ".";
```

You haven't seen the (led.Read() ? "ON" : "OFF") construct before. Here's how it works: the statement before the ? is evaluated as true or false. If it is true (the LED is ON), then the value after the question mark ("ON") is chosen. Otherwise, the value after the colon is chosen ("OFF").

Now you can create a web page to return to the client:

```
// return a message to the client letting it
// know if the LED is now on or off.
string response =
  "HTTP/1.1 200 OK\r\n" +
  "Content-Type: text/html; charset=utf-8\r\n\r\n" +
  "<html><head><title>Netduino Plus LED Sample</title></head>" +
  "<body>" + statusText + "</body></html>";
```

This is a pretty basic web page. It will simply let the user know whether the LED is now ON or OFF, usually in response to their request. I've included the **statusText** variable from earlier.

Now send the response on its way:

```
clientSocket.Send(System.Text.Encoding.UTF8.GetBytes(response));
```

There's one last piece of business to take care of. After the next curly brace (which ends the if block), you should close the client socket connection:

```
}

// important: close the client socket
clientSocket.Close();
```

Your full code will look like this:

```
public static void Main()
{
    // write your code here

    // setup the LED and turn it off by default
    OutputPort led = new OutputPort(Pins.ONBOARD_LED, false);

    // configure the port # (the standard web server port is 80)
    int port = 80;

    // wait a few seconds for the Netduino Plus to get a network address.
    Thread.Sleep(5000);
```

```csharp
// display the IP address
Microsoft.SPOT.Net.NetworkInformation.NetworkInterface
  networkInterface =
    Microsoft.SPOT.Net.NetworkInformation.NetworkInterface.
      GetAllNetworkInterfaces()[0];

Debug.Print("my ip address: " + networkInterface.IPAddress.ToString());

// create a socket to listen for incoming connections
Socket listenerSocket = new Socket(AddressFamily.InterNetwork,
                                   SocketType.Stream,
                                   ProtocolType.Tcp);
IPEndPoint listenerEndPoint = new IPEndPoint(IPAddress.Any, port);

// bind to the listening socket
listenerSocket.Bind(listenerEndPoint);
// and start listening for incoming connections
listenerSocket.Listen(1);

// listen for and process incoming requests
while (true)
{
    // wait for a client to connect
    Socket clientSocket = listenerSocket.Accept();

    // wait for data to arrive
    bool dataReady = clientSocket.Poll(5000000, SelectMode.SelectRead);

    // if dataReady is true and there are bytes available to read,
    // then you have a good connection.
    if (dataReady && clientSocket.Available > 0)
    {
        byte[] buffer = new byte[clientSocket.Available];
        int bytesRead = clientSocket.Receive(buffer);

        string request =
          new string(System.Text.Encoding.UTF8.GetChars(buffer));
        if (request.IndexOf("ON") >= 0)
        {
            led.Write(true);
        }
        else if (request.IndexOf("OFF") >= 0)
        {
            led.Write(false);
        }

        string statusText = "LED is " + (led.Read() ? "ON" : "OFF") + ".";
```

```
    // return a message to the client letting it
    // know if the LED is now on or off.
    string response =
        "HTTP/1.1 200 OK\r\n" +
        "Content-Type: text/html; charset=utf-8\r\n\r\n" +
        "<html><head><title>Netduino Plus LED Sample</title></head>" +
        "<body>" + statusText + "</body></html>";
    clientSocket.Send(System.Text.Encoding.UTF8.GetBytes(response));
    }

    // important: close the client socket
    clientSocket.Close();
    }
}
```

Now, go ahead and run your app.

After about five seconds, you should see the following text in Visual Studio's Output window:

```
my ip address: 192.168.5.133
```

 NOTE: Your IP address will vary; this is the one that was given to my Netduino Plus by my router.

Using that address, use a web browser on your local network to open up a connection to your Netduino Plus. If you have a portable device (phone, tablet, etc.) connected to your WiFi network, that will work too.

Open up the following address in your browser, replacing *192.168.5.133* with your Netduino's IP address (which is displayed in the Output window; to display this window, see Summoning the Output Window on page 66): *http://192.168.5.133*.

You will see a web page that says:

```
LED is OFF.
```

Now, change the web page request to include the word "ON." Again, substitute your Netduino's IP address:

```
http://192.168.5.133/ON
```

The LED on your Netduino Plus should turn on, as shown in Figure 8-1! And the response should read:

```
LED is ON.
```

Figure 8-1. *Turning on and off the LED on the Netduino Plus—via the Internet*

By changing the word ON in the web page request, you can also turn your Netduino Plus's LED off again:

```
http://192.168.5.133/OFF
```

Conclusion

You have now created your first Internet-enabled Netduino app. Along with the other skills you have acquired during this book, you can create sensors that tweet. Or robots that move in response to web-based navigation commands. The possibilities are endless.

A/Upgrading Firmware

One of the founding principles of Netduino is that Netduino's hardware and software are open source. Because of this it's possible to compile the latest Netduino firmware from scratch and reflash your Netduino, to create custom features and incorporate them into the firmware, and to share those custom features with others. Netduino is designed to be fully "hackable."

The Netduino firmware is made up of two pieces of software: the bootloader (TinyBooter) and the .NET Micro Framework runtime (TinyCLR).

Each variant of Netduino (Netduino, Netduino Plus, and Netduino Mini) has separate versions of the firmware. Make sure you have the right firmware for the board you want to flash. New firmware releases are posted to *http://forums.netduino.com* under the Hardware section. Go into the subsection (Netduino, Netduino Plus, or Netduino Mini) for the board you want to flash, and look for the posts at the top of the list (they'll have a title like "Netduino Firmware" followed by a version and update number).

The bootloader runs when the board is first powered on. Its jobs are to load the .NET MF (Micro Framework) runtime and to enable you to update the .NET MF runtime. The .NET MF runtime then runs your Netduino app.

Due to the bonded nature of these two pieces of software, both should be updated whenever a new release of the .NET Micro Framework is released. For minor updates and patches, only the .NET MF runtime needs to be reflashed. Updates are optional and only take a minute or two.

Minor Updates

For minor updates, use the MFDeploy tool, which ships with the .NET Micro Framework SDK to update the .NET MF runtime on the Netduino.

To begin, run the MFDeploy application. You can find this in Start→All Programs→.NET Micro Framework SDK→Tools→*MFDeploy.exe*.

NOTE: If you have file extensions hidden in Windows Explorer, several files may appear to be named *MFDeploy* or *MFDeploy.exe*. Make sure you open the one that's an application.

In MFDeploy, change the deployment method from Serial to USB and ensure that your Netduino appears in the drop-down box. Now, press Ping (see Figure A-1) to ensure that your device is available and responding.

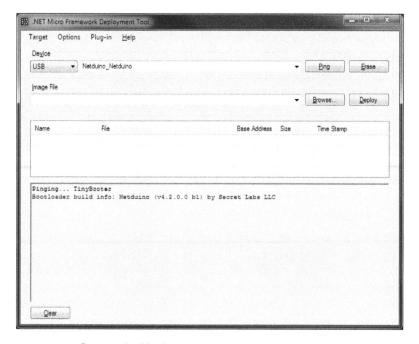

Figure A-1. *Pinging the Netduino*

NOTE: If your Netduino doesn't respond, press the Netduino's SW1 pushbutton. Then unplug and reconnect your Netduino to your PC. Press Ping on your PC to see "Pinging... TinyBooter" and then release the pushbutton on your Netduino. This failsafe mode will put your Netduino into a special bootloader mode for 20 seconds so you can erase your app or update the firmware.

In the Image File section, press the Browse button to select the updated Netduino firmware. The two files will be named *ER_CONFIG* and *ER_FLASH*. Browse for these and press OK once you have selected them.

Now, press Deploy. This will reflash your Netduino's .NET MF firmware and may take a minute (see Figure A-2).

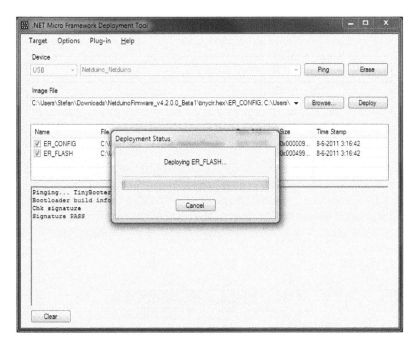

Figure A-2. *Deploying firmware*

Once deployment has completed, verify that your firmware has been updated to the newest version. You can do this by selecting Device Capabilities in the Target menu (or pressing Ctrl+Shift+C).

The output will contain a SolutionReleaseInfo.solutionVersion and Solu tionReleaseInfo.solutionVendorInfo. They should look something like this:

```
SolutionReleaseInfo.solutionVersion:    4.1.0.0
SolutionReleaseInfo.solutionVendorInfo: Netduino (v4.1.0.0) by Secret Labs
LLC
```

Major Upgrades

When moving between new releases like .NET Micro Framework 4.1 and .NET Micro Framework 4.2, you should update the Netduino's bootloader as well.

To do so, use Atmel's "SAM-BA CDC" software (current version is 2.11). If you don't already have a copy, download it now.

 NOTE: At the time of writing, this tool can be found at *http://www.atmel.com/dyn/products/tools _card.asp?tool_id=3759*. Up-to-date information on reflashing Netduino can be found in the Netduino community at *http://wiki.netduino.com*.

To reflash the bootloader, you need to erase everything from the Netduino's microcontroller flash memory. To do so, make sure your Netduino is powered up, then connect a wire between the 3.3V header on the Netduino and the small gold square directly below digital pin D0 as shown in Figure A-3. Applying power to this pad for half a second will erase the Netduino's firmware.

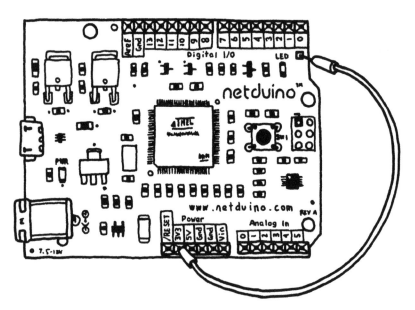

Figure A-3. *Applying 3.3V power to the erase pad*

Now, unplug and reconnect your Netduino. Windows should detect a new serial port, which is a virtual serial port that will enable you to update the bootloader.

Now start the SAM-BA CDC tool (Figure A-4). Select the COM port assigned to your Netduino and select a board type of at91sam7x512-ek. Press Connect.

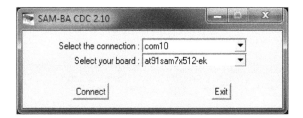

Figure A-4. *Connecting to Netduino*

 NOTE: If there are multiple COM ports available and you're not sure which one is the virtual one belonging to your Netduino, open up Windows Device Manager and look under the Ports category.

Once SAM-BA CDC connects to your Netduino, you need to set it up to boot from flash. Select Boot from Flash (GPNVM2) from the scripts drop-down and press Execute, as shown in Figure A-5.

Then select the Enable Flash Access script and press Execute.

Once these are complete, you can flash the TinyBooter bootloader to the Netduino. Click the Open button to the right of the Send File Name field and select the *TinyBooterDecompressor.bin* file for your Netduino firmware. Then, press Send File.

After a few moments, SAM-BA will complete the flashing process and will ask if you want to lock regions. Click No.

Optionally, you can verify that the bootloader was flashed properly. To do so, increase the value of the "size (for receive file)" field (adding 2-3 zeros at the end is plenty) and then press "Compare sent file with memory." The file should match!

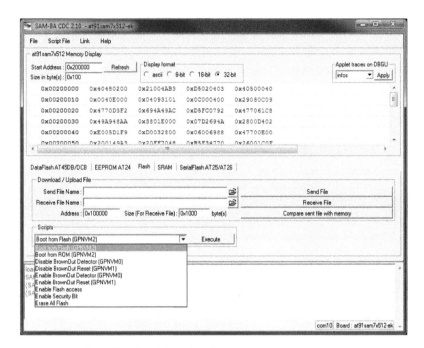

Figure A-5. *Selecting Boot from Flash*

Close SAM-BA and disconnect and reconnect your Netduino. The new boot-loader will start up and the Netduino is now ready to have the .NET Micro Framework runtime flashed. Follow the instructions in the section "Minor Updates" on page 73 to complete the flashing process.

B/Developing Netduino Apps with Mono

Although Visual Studio is the fully supported way to develop for Netduino, there is an alternative that allows you to develop Netduino apps on Mac OS X and Linux: Mono, an open source implementation of the .NET Framework CLR. Mono doesn't include the Micro Framework, but if you're comfortable with your system's command line, then you can take the open source bits of the .NET Micro Framework and integrate them with your own installation of Mono.

NOTE: Because you'll be loading compiled code from a flash memory card, this works best with a Netduino Plus (because it has an integrated Micro SD card slot). However, you can attach a MicroSD breakout board such as Adafruit's MicroSD card breakout board+ (*http://www.adafruit.com/prod ucts/254*) to a Netduino or Netduino Mini and use that.

Also, you'll need to have access to a Windows machine just once, to install a special bootstrapping app on your Netduino that allows you to load programs from the MicroSD card.

Prepare the Netduino

For this step, you'll need to use a Windows computer and install Visual Studio on it to deploy an app to your Netduino. Chapters 2 and 3 have all the information you need to do this. Once you're able to run apps on your Netduino, download *MonoNetduinoBootloader.zip* from *http://examples.oreilly.com/ 0636920018032/*. Unzip the file and open the project in Visual Studio. Go into the project properties, set the deployment target to your Netduino, and click Debug→Start Debugging. The app will not run successfully because you don't (yet) have a Netduino app on the MicroSD card. Power down the Netduino for now (you can just unplug it from the computer).

Set Up Your Developer Tools

First, you'll need to make sure you've got development tools installed on your machine. On Mac OS X, you should download and install Xcode from *http://developer.apple.com/xcode*. If you're on Linux, you'll need to install all the packages you need for C development, as well as a few other tools (autoconf, automake, libtool, bison, flex, and gettext). On Ubuntu Linux, for example, you can install these with the following **apt-get** command at the terminal (you'll be prompted for your password whenever you run a command with sudo):

```
sudo apt-get install build-essential autoconf \
    libtool bison flex gettext
```

Install Mono

At the time of this writing, the current release version of Mono (2.10) did not support Netduino. So you'll need to compile the leading edge version (2.12) of Mono using the current release version.

Install Mono 2.10

Head over to the Mono download page at *http://www.go-mono.com/mono-downloads/download.html* and follow the instructions to install it for your operating system. If you get there and the version of Mono is 2.12 or later, you can skip the next step. On Linux, you should be able to install it using your package manager. On Ubuntu Linux, for example, you can install the minimum Mono infrastructure needed by running this **apt-get** command at the terminal:

```
sudo apt-get install mono-mcs mono-gmcs
```

Compile Mono 2.12 with Mono 2.10

If you installed Mono 2.10 in the previous step, you will need to compile Mono from source. The easiest way to download it is with Git, a version control system. Mac OS X Lion includes Git, but if you need to download it for an older version of Mac OS X, go to *http://git-scm.com*. On Linux, you should install the **git** package. On Ubuntu Linux, for example, you can install it by running this **apt-get** command at the terminal:

```
sudo apt-get install git
```

Now that you have Git installed, open up a terminal window, change directory to wherever you want to download the source to, and issue this command:

```
git clone git://github.com/mono/mono.git
```

This will create a subdirectory named *mono*. Next, change directory into that subdirectory:

```
cd mono
```

Then, configure the source tree. Use this command on Mac OS X:

```
./autogen.sh --prefix=/usr/local --with-glib=embedded --enable-nls=no
```

And this command on Linux:

```
./autogen.sh --prefix=/usr/local
```

Now you can use the make command to build mono:

```
make
```

If it compiles successfully, you can install it with this command (you'll be prompted for your password when you run it):

```
sudo make install
```

Now you have mono installed in */usr/local/bin*. You also have an older version installed in */usr/bin*, but you won't use that for Netduino development.

If you have any trouble compiling Mono, see *http://www.mono-project .com/Compiling_Mono_From_Git*.

Install Wine

All but one of the .NET Micro Framework tools you need to build apps are .NET apps, and run just fine under Mono. *MetaDataProcessor.exe* is a native Windows app, but you can run it under Wine, an implementation of Windows APIs that runs on Mac OS X or Linux. You'll first need to install Wine (see *http://www.winehq.org/* for details). On Linux, you should be able to install it using your package manager. For example, on Ubuntu, you can use this command at the terminal:

```
sudo apt-get install wine
```

After you've installed Wine, run `winecfg` from the terminal window to get Wine set up for the first time. As soon as the Wine Configuration window appears, you can close it.

Next, you need to add some libraries. The fastest way to do this is by installing the *winetricks* utility. See *http://wiki.winehq.org/winetricks* for information on installing and using it (if you installed Wine with a package manager on Linux, it may have installed *winetricks* along with Wine). After you install it, use it to install these libraries and packages (each one may cause one or more installation dialogs to appear):

- vcrun2010
- vcrun2008
- dotnet30

Download Supporting Binaries

You're almost there. Next, you need the *netmfbins.zip* archive that contains several files from the open source .NET Micro Framework needed to compile Netduino apps. You can find this file at *http://examples.oreilly.com/0636920018032/*. Create a directory that you'll use to compile your Netduino app. Unzip the *netmfbins.zip* file and put its contents into that directory. This zip file also contains a *makefile*, which the command-line make utility uses to compile programs. You can use this makefile as-is, or customize it for your needs.

Compile an App

Now you've got a directory with the following files in it:

- Makefile
- MetaDataProcessor.exe
- Microsoft.SPOT.Hardware.dll
- Microsoft.SPOT.Native.dll
- Microsoft.SPOT.TinyCore.dll
- SecretLabs.NETMF.Hardware.Netduino.dll
- SecretLabs.NETMF.Hardware.dll
- mscorlib.dll

All you need to add is a program (named *Program.cs*) and a subdirectory (*Properties*) with a file called *AssemblyInfo.cs* in it.

Try it out with the program from Chapter 3. Here's what needs to go inside your *Program.cs*:

```
using System;
using System.Threading;
using Microsoft.SPOT;
using Microsoft.SPOT.Hardware;
using SecretLabs.NETMF.Hardware;
using SecretLabs.NETMF.Hardware.Netduino;

namespace MonoNetduinoApp
{
    public class Program
    {
        public static void Main()
        {
            // write your code here
            OutputPort led = new OutputPort(Pins.ONBOARD_LED, false);

            while (true)
            {
                led.Write(true); // turn on the LED
                Thread.Sleep(250); // sleep for 250ms
                led.Write(false); // turn off the LED
                Thread.Sleep(250); // sleep for 250ms
            }
        }
    }
}
```

And here's what you need inside of *Properties\AssemblyInfo.cs* (you will need to create the *Properties* subdirectory):

```
using System.Reflection;
using System.Runtime.CompilerServices;
using System.Runtime.InteropServices;

[assembly: AssemblyTitle("Blinky")]
[assembly: AssemblyDescription("")]
[assembly: AssemblyConfiguration("")]
[assembly: AssemblyCompany("")]
[assembly: AssemblyProduct("Blinky")]
[assembly: AssemblyCopyright("")]
[assembly: AssemblyTrademark("")]
[assembly: AssemblyCulture("")]

[assembly: AssemblyVersion("1.0.0.0")]
[assembly: AssemblyFileVersion("1.0.0.0")]
```

With all that in place, type the command **make** at the terminal. You should see output similar to the following:

```
/usr/local/bin/mcs -nostdlib -target:library -out:MonoNetduinoApp.dll
Program.cs Properties/AssemblyInfo.cs -
r:SecretLabs.NETMF.Hardware.Netduino.dll
-r:Microsoft.SPOT.TinyCore.dll -r:Microsoft.SPOT.Hardware.dll
-r:Microsoft.SPOT.Native.dll -r:mscorlib.dll

wine MetaDataProcessor.exe -loadHints mscorlib mscorlib.dll -parse
MonoNetduinoApp.dll -minimize -endian le -compile MonoNetduinoApp.pe
```

When you're done, you can copy the file *MonoNetduinoApp.pe* to the root of your memory card, put it in the Netduino, and power it up. It should start running the Netduino app you compiled on Mac OS X or Linux. If you have any troubles with this, check out the Mono forum (*http://forums.netduino .com/index.php?/forum/12-mono/*) at *http://forums.netduino.com/* and post a question if you can't find your answer there.

About the Author

Chris Walker is the inventor of the Netduino, host of the Netduino user community, and an expert on .NET Micro Framework.

Have it your way.

Get even more for your money.

Join the O'Reilly Community, and register the O'Reilly books you own. It's free, and you'll get:

- $4.99 ebook upgrade offer
- 40% upgrade offer on O'Reilly print books
- Membership discounts on books and events
- Free lifetime updates to ebooks and videos
- Multiple ebook formats, DRM FREE
- Participation in the O'Reilly community
- Newsletters
- Account management
- 100% Satisfaction Guarantee

Signing up is easy:

1. Go to: oreilly.com/go/register
2. Create an O'Reilly login.
3. Provide your address.
4. Register your books.

Note: English-language books only

To order books online:

oreilly.com/store

For questions about products or an order:

orders@oreilly.com

To sign up to get topic-specific email announcements and/or news about upcoming books, conferences, special offers, and new technologies:

elists@oreilly.com

For technical questions about book content:

booktech@oreilly.com

To submit new book proposals to our editors:

proposals@oreilly.com

O'Reilly books are available in multiple DRM-free ebook formats. For more information:

oreilly.com/ebooks

O'REILLY®